MW01491685

ELECTRICIAN'S BOOK

CONTROL CIRCUITS

By CORNEL BARBU

Copyright 2007 by Cornel Barbu with registration

ISBN:978-1-4357-0782-5

ID#1023533

www.lulu.com

Dedicated to Mike and Florin

"So you never forget that there is solution for every situation you are going through.

Stay calm and think..."

Desperation is lack of thinking...........

CONTENTS

WHY I AM WRITING THIS BOOK

One of the reasons I am writing this book is so you can learn from my experience and successfully achieve your tasks as an electrician. If you are someone who dreams big, I can help you succeed in the construction business- the most exciting and fulfilling career you can choose!

After reading this book, you will be able to competently and confidently perform electrical tasks as an electrical apprentice or electrician. You can then proceed to the next level as a leader in this field if you want to do so. If you choose this path and wish go into business for yourself, please refer to my book:

"Close to my first million dollars as an electrical contractor" ISBN: 978-1-4303-1969-6

If you want to stay as a professional electrician you'll have all the information necessary to be successful in North America. It is a place where construction is booming and because of technological advances there are changes and upgrades needed everywhere. The field is endless and work is abundant in the industrial, commercial, agriculture, mining and residential sectors. Rest assured you are in the right field!

I am writing these books to try and help those individuals who are passionate about this field and wish to build a career as electrician or run their own business as an electrical contractor.

A series of ELECTRICIAN'S BOOKS with different applications are in development. These will cover all aspects of the business and provide a straightforward, consistent source of information to you. These will be available in either hard copy or for downloading from online stores. Some of titles will be:

1. ELECTRICIAN'S BOOK –CONTROL CIRCUITS
2. ELECTRICIAN'S BOOK -CONDUIT INSTALLATION
3. ELECTRICIAN'S BOOK -FIRE ALARM SYSTEM INSTALLATION
4. ELECTRICIAN'S BOOK -WIRING AND CABLE INSTALLATION

5. ELECTRICIAN'S BOOK -HOW TO PRICE ELECTRICAL WORK
6. ELECTRICIAN'S BOOK -EXTERIOR ELECTRICAL WORK
7. ELECTRICIAN'S BOOK -HOW TO READ ELECTRICAL DRAWINGS
8. ELECTRICIAN'S BOOK -MATEHMATICS FOR ELECTRICIANS
9. EXAM PREPARATION FOR CANADIANS
10. ELECTRICIAN IN NORTH AMERICA

There is a secret to securing your future.

Competence!

COMPETENCE = KNOWLEDGE + EXPERIENCE

I wish you the best!

CONTROL DIAGRAM

A control diagram uses symbols to indicate the sequence of specific tasks in a logical way. Tasks such as connecting or disconnecting the power for certain devices or interlocking different electrical devices are explained by control diagrams.

For electricians, control diagrams can be one of the most challenging aspects of the job since they require us to be able to make careful observations and think things through in a logical way. They require us to consciously process our understanding. Being able to follow and create control diagrams means we can observe how things are working inside of a technological process and troubleshoot when required. Your ability to repair and maintain installations using a control diagram is proof of your expertise and skill as an electrician. It will secure your position as electrician in any organization!

This book will begin by providing you with the basics and then build upon this to give you a thorough understanding and training in control circuitry. Let's begin! One of the most simple and common examples of control diagrams are those for lighting systems. In these systems there are two types of controls:

• Manual control

• Auto control system

When in manual control the user will control the electrical flow or circuit by closing or opening it manually. For example, we can turn lights ON and OFF by activating a lighting switch located on a wall or other device, such as a lighting fixture.

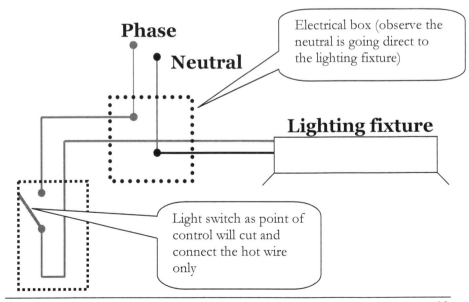

Phase

Neutral

Electrical box (observe the neutral is going direct to the lighting fixture)

Lighting fixture

Light switch as point of control will cut and connect the hot wire only

Another example can be found when activating a pump. We can start a pump motor in the "manual" mode from the starter by simply pushing a button (PB is ON). And we can stop the pump by pressing the stop button (Off).

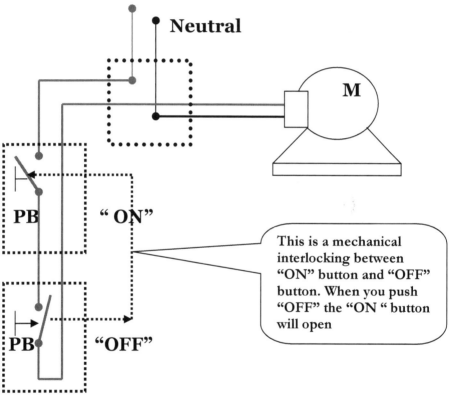

In other words, to supply the motor with electrical power, you need to push the "ON" button. To disconnect or stop this power flow, you need to push the "OFF" button. In some cases this will be done by automated or logic devices that initiate the task based on a diagram or sequence. *Automatic systems* do not rely on human action. An example of an automatic system is one in which lights will turn ON or OFF depending on the presence of people occupying an area requiring lighting. These systems are usually dependant on sensors and/or timers.

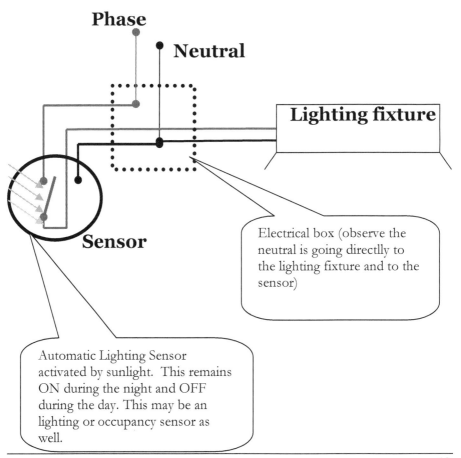

Phase

Neutral

Lighting fixture

Sensor

Electrical box (observe the neutral is going directlly to the lighting fixture and to the sensor)

Automatic Lighting Sensor activated by sunlight. This remains ON during the night and OFF during the day. This may be an lighting or occupancy sensor as well.

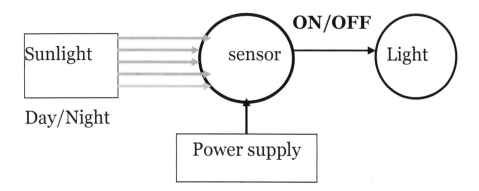

This is the bloc diagram. The sunlight will trigger the sensor and the sensor will switch the power ON or OFF for the lighting fixture. In this case the sensor is able to connect and disconnect the load form power supply (where the lighting fixture is the load).In such situations where the load is appreciable will be required to install a contactor that will be controlled by the sensor. The sensor will energize or de-energize the contactor that supplies the power to the lighting system. Obviously with this type of sensor we can control all sort of other loads.

For example, a sensor can be used to close or open window blinds. With imagination and common sense, the possibilities are endless!

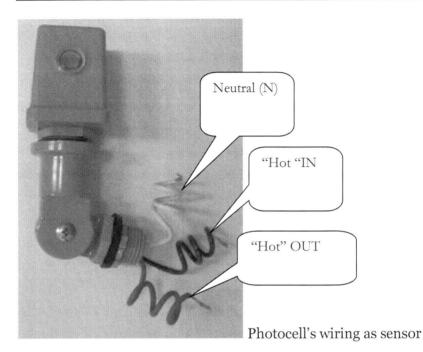

Neutral (N)

"Hot "IN

"Hot" OUT

Photocell's wiring as sensor

This connector will make an installation possible to any junction box for wiring terminations

Observe the rubber gasket which ensures waterproof conditions for the junction box and terminations

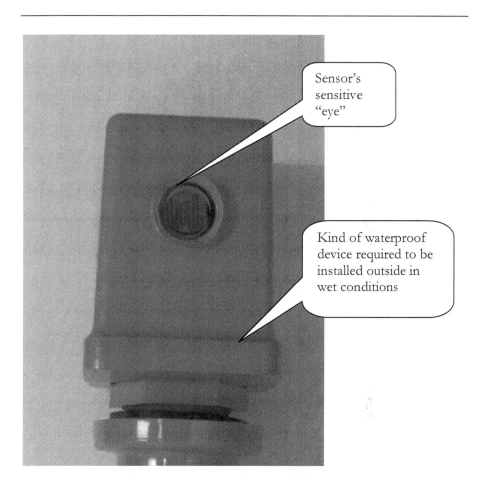

Your notes here:

SYMBOLS FOR POWER/CONTROL DIAGRAM

When information needs to be transferred between individuals we generally use oral language or words. Our linguistic skills are critical in allowing us to communicate with each other. In the electrical trade there is a *common language and terminology* used to help electricians communicate with one another.

Most of the time we use drawings to communicate the necessary information required to complete an electrical installation.

It is the responsibility of the designers to prepare all of the documentation to go with their design. This means they will provide trade drawings, specifications and notes in order to make their "vision" or design easy to understand.

When designers provide the drawings they use symbols for electrical devices. As electricians we need to be able to recognize all these symbols in order to properly read the drawings. Most drawings will also include notes and a legend to identify and describe each electrical symbol used in that drawing. As you start your career as an apprentice and electrician you will need to refer to these notes and legends. Over time you'll learn the "language" of these symbols. Reading and recognizing the symbols will become progressively more familiar and easier with each job.

You will accumulate the experience and knowledge necessary to secure a steady flow of job opportunities. Over the next few pages you will learn to recognize the electrical symbols for control diagrams.

Due to the fact that control diagrams are linked with the power diagrams, some of the power symbols used in electrical diagrams will also be included in this section.

In control diagrams most of the time we find *relays, contacts, indicating lights, push buttons and sensors.* There are other symbols that haven't been included here and during the course of your work you'll discover and learn about a variety of different symbols.

"NO" stands for: normally open contact
"NC" stands for: normally closed contact

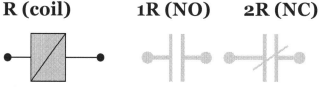

R (coil)　　　**1R (NO)**　　**2R (NC)**

1R is the first contact NO of relay R since 2R is the second contact NC of relay R

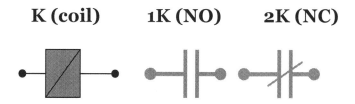

K (coil)　　　**1K (NO)**　　　**2K (NC)**

1K is the first contact NO of contactor K since 2K is the second contact NC of contactor K

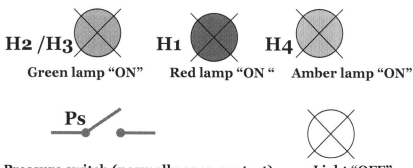

H2 /H3 Green lamp "ON"　　**H1** Red lamp "ON "　　**H4** Amber lamp "ON"

Ps Pressure switch (normally open contact)　　Light "OFF"

PB (push button NO)   **PB (push button NC)**

In power diagrams we find *fuses, power supply sources, power contacts, contactors, transformers, overload relays, motors and means of disconnecting power*. Grounding or bonding wire and bus bars representation will appear also in power diagrams. The illustration below shows these parts of the power circuit.

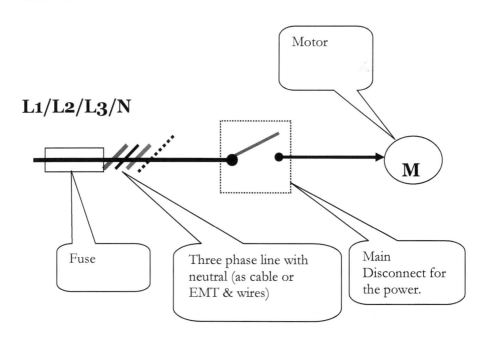

L1-phase 1/L2-phase2/L3-phase3/N-neutral

Power diagram & symbols:

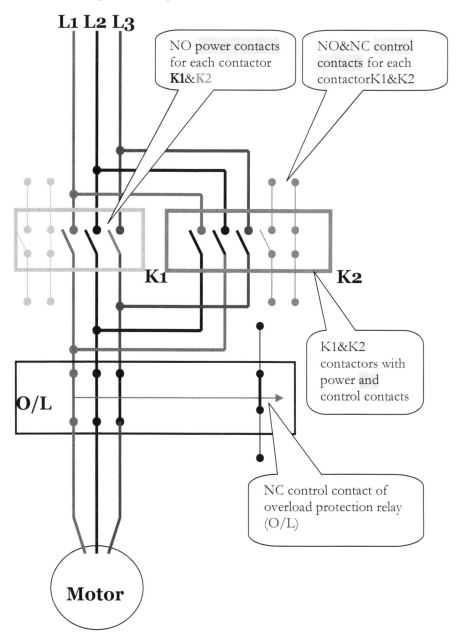

L1 L2 L3

NO power contacts for each contactor **K1**&K2

NO&NC control contacts for each contactorK1&K2

K1   K2

O/L

K1&K2 contactors with power and control contacts

NC control contact of overload protection relay (O/L)

Motor

THE BLOC DIAGRAM

In order to understand the control circuit for different devices or equipment you need to understand the bloc diagram for that device.

The bloc diagram will include:

1. Parameters
2. Logic diagram
3. Controlled device
4. Power supply

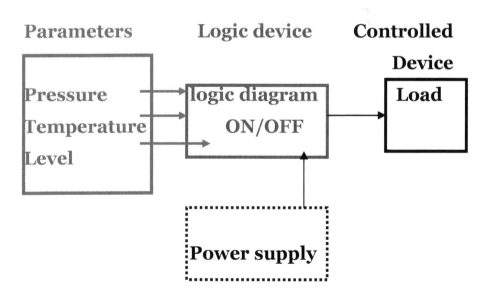

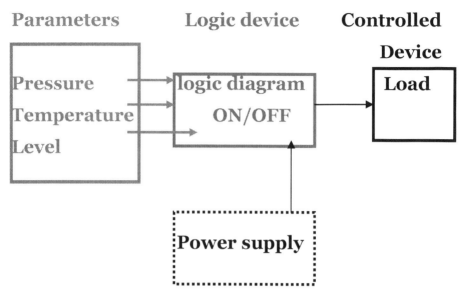

Parameters

Variables or parameters such as temperature, levels, currents, voltage and pressure almost always need to be taken into consideration as they trigger the logic diagram.

The parameters will trigger the logic diagram and as a result the load will be switched ON or OFF to the power supply.

So before you attempt to troubleshoot using a control diagram, you need to know how the process is working. This requires knowing what the parameters or variables are that will trigger the logic diagram and how the logic diagram has been designed and works in order to control the load.

Devices controlling parameters such as pressure or level will provide a contact for that parameter in the logic diagram. These contacts will be NO (normally open) or NC (normally closed).

LOGIC DIAGRAM

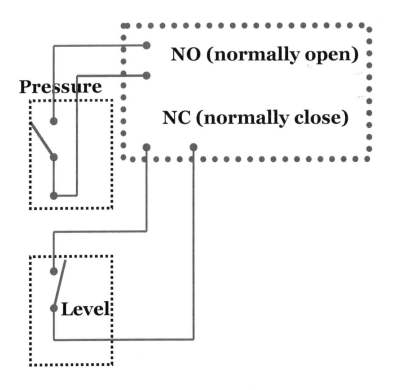

Logic diagram Air tank /pressure control circuit (example)

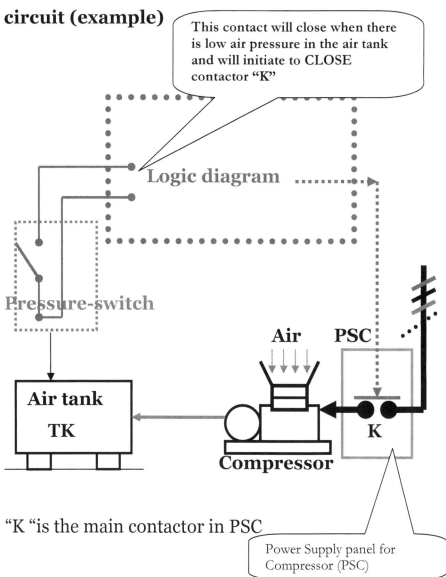

This contact will close when there is low air pressure in the air tank and will initiate to CLOSE contactor "K"

Logic diagram

Pressure-switch

Air   PSC

Air tank
TK

K

Compressor

"K "is the main contactor in PSC

Power Supply panel for Compressor (PSC)

Later on we will study the control circuit for this type of a logic diagram.

Mechanical or electronic devices will control the parameters and will provide the contacts (NO or NC). These will be integrated into the logic diagram and will have specific functions.

In the previous air tank example, the pressure switch located on the tank provides a NO contact that <u>CLOSES</u> when the pressure of the air in the tank reaches a low level. The logic diagram uses this contact to energize a contactor that supplies power for a compressor. This compressor then increases the air pressure in the tank.

When the air pressure is once again within the prescribed limits, the pressure switch <u>will OPEN</u> and cause the logic diagram to disconnect the power from the compressor.

At this point we are not going to go into detail about the logic diagram as the main goal of this section is simply to understand how the basic process works. Later on we will do a more in-depth analysis of each part of a control circuit diagram.

Controlled device

In the aforementioned case, the controlled device is the compressor which is the air supply source for the air tank. The pressure switch keeps the air pressure within the limits set for the tank.

Power supply panel (PSC)

This device is able to provide the necessary current, voltage and frequency for the compressor. The contactor K will close when all the conditions of the logic diagram are in place and will energize the compressor. Control circuits that are part of the logic diagram will be provided with a power supply as well. This is usually independent from the power distribution circuit. We will now identify (page 24) these two different types of circuits:

1.   Power circuit (the highlighted line)
2.   Control circuit (dotted lines)

CONTROL RELAY

This device will be found in many applications and it is critical that you know some basic information regarding the relay's operation and its main parts.

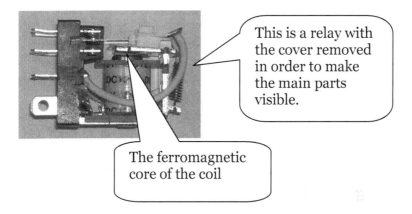

This is a relay with the cover removed in order to make the main parts visible.

The ferromagnetic core of the coil

The main part of the relay is the coil. This coil has a ferromagnetic core that magnetizes when it is supplied with power, or an electrical voltage, at its terminals. In most control diagrams the symbol for the coil will look like the diagram below:

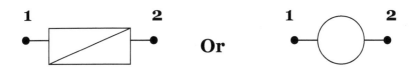

Or

Applying voltage (power) to terminals 1&2 will cause a current through the coil. This current will then create a magnetic flux which magnetizes the ferromagnetic core of the coil. The next diagram shows the location of the coil, core and contacts within the main parts of the relay. Make your notes beside the picture or into empty area of the page.

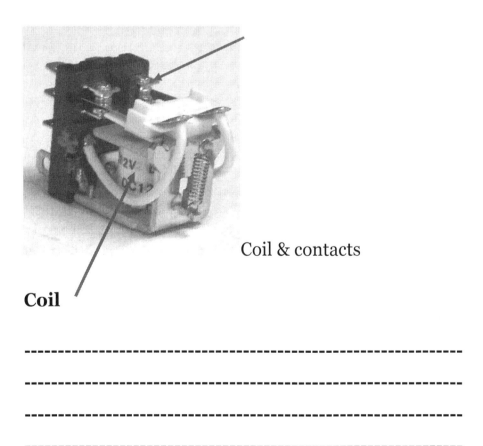

Coil & contacts

Coil

Back plate with terminals

#5 NC

#4NO

#3 common

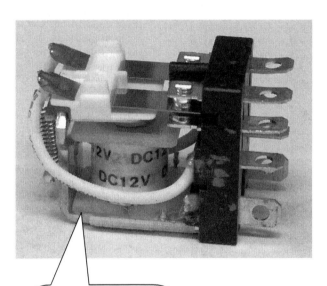

These relays operate at 12 volts DC as indicated on the coil's sticker

The relay's main parts (drawings):

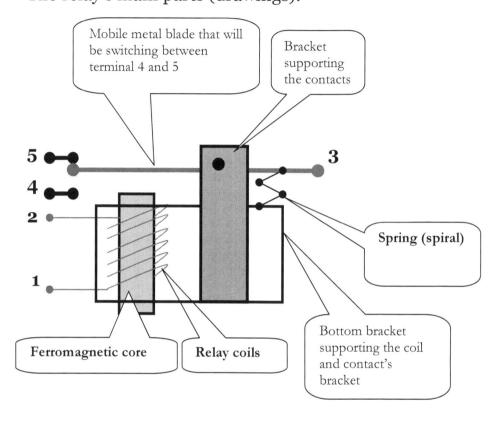

Schematics of the contacts:

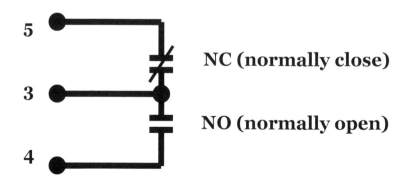

NC (normally close)

NO (normally open)

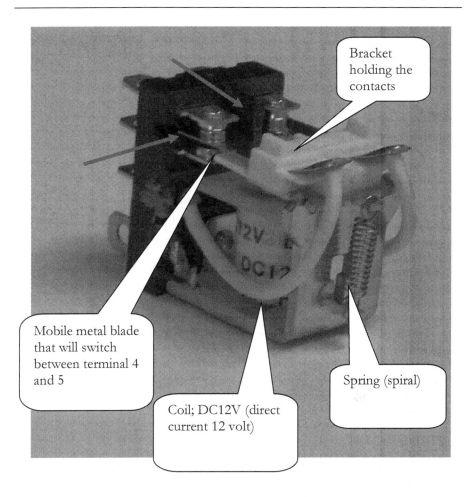

Bracket holding the contacts

Mobile metal blade that will switch between terminal 4 and 5

Spring (spiral)

Coil; DC12V (direct current 12 volt)

The ferromagnetic core will not be visible in this picture since is the core of the coil. Also, please observe the contacts of the relay.

These are indicated with arrows. In this picture the protection cover of the relay been removed to make it easier to see inside the relay.

The relay's terminals are more visible in this photo. Please make your notes here as well!

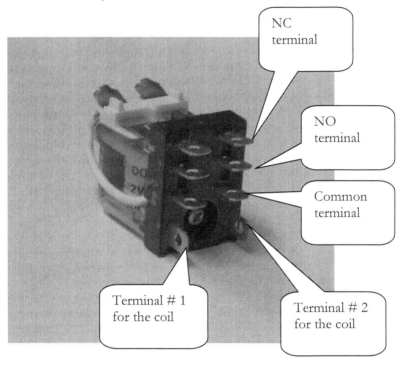

Since the core will be magnetized it will attract the metal blade and open the electrical contact between 3 and 5. In other words, the normally closed contact will open. The blade will then switch to contact 4 so that the contact between 3 and 4 will be close (see page 30). This means the contact that is normally open is now closed.

In short, this means when a voltage is applied to the coil at terminals 1 and 2, the normal status of the relay's contact will change.

Now that you have a basic understanding of how the relay works, we will explore the way in which a relay is integrated and functions within a control diagram (See page 42).

"Repetitio mater studiorum est." The English translation of this Latin expression means that repetition is the mother of study, so please be sure to have a notebook while you read this. It is important that you make notes while you read! You should often review these notes to refresh your memory about facts and knowledge that will help you succeed as an electrician.

The previous part of this book introduced the terminology that is the basis of an electrician's day-to-day technical language. We will now use this previous information and apply it to understand the next illustration:

Relay status for de-energized coil

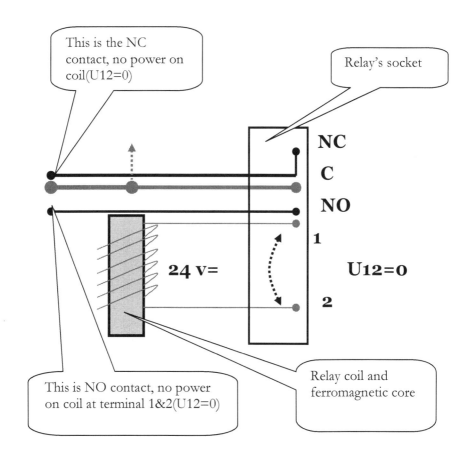

The relay's main parts are:

- Ferromagnetic core
- Coil
- Relay's socket
- A pair of NC and NO contacts

For your review

Ferromagnetic core

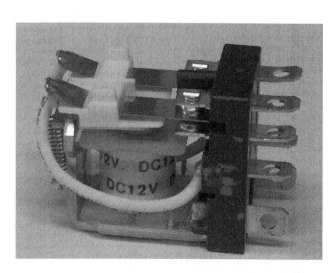

This device works in the following way:

When the coil is energized, the ferromagnetic core magnetizes and changes the status of the contacts.

The NC (normally closed) contact will OPEN and

The NO (normally open) contact will CLOSE.

The voltage rating for a coil varies from 12 volts to 120 volts. Relays work in either DC or AC currents. Since the main part of the relay is the coil and ferromagnetic core, it is important to note that there are differences between AC and DC relays. A DC relay will have a ferromagnetic core that is a cylindrical solid metal core. An AC relay will have a ferromagnetic core that is segmented or looks like a series of strands. The diagram below illustrates this difference:

DC ferromagnetic core (cross section)

"Sandwich "of ferromagnetic plates to build the magnetic core

The next diagram is a representation of a relay
mechanism. The main parts of the relay are indicated.

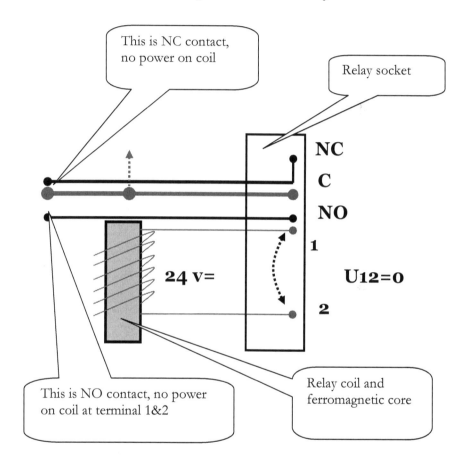

In this diagram, the relay's coil is not energized since
U12=0(coil terminals 1&2).

This next example shows a DC relay with a coil working at 24 volts DC.

ENERGIZED COIL situation:

Power at terminal 1 & 2 U12 =24V DC

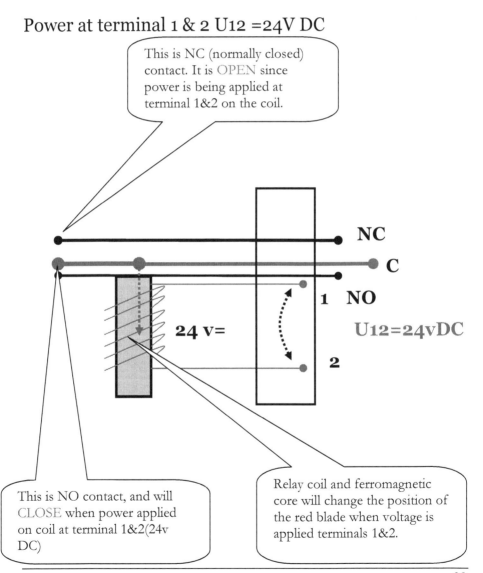

This is NC (normally closed) contact. It is OPEN since power is being applied at terminal 1&2 on the coil.

NC

C

1 NO

24 v=

U12=24vDC

2

This is NO contact, and will CLOSE when power applied on coil at terminal 1&2(24v DC)

Relay coil and ferromagnetic core will change the position of the red blade when voltage is applied terminals 1&2.

So, if the relay is not energized:

Contact 3-4 is   NO (normally open)

Contact 3-5 is NC (normally close)

Uc =coil voltage (U12) =0

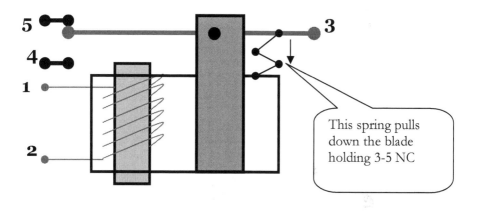

This spring pulls down the blade holding 3-5 NC

The schematic of the NO &NC contact is:

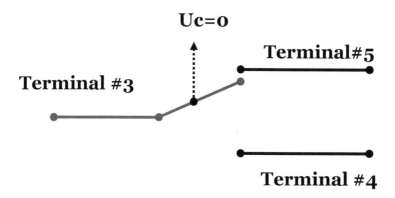

Uc=0

Terminal #3

Terminal#5

Terminal #4

If the relay <u>is energized</u> then Uc =U12=24 v DC and

NO contact 3-4 will close

NC contact 3-5 will open.

U coil =24 volt DC.

Electromagnetic force directions able to change the blade positions so the contact's status.

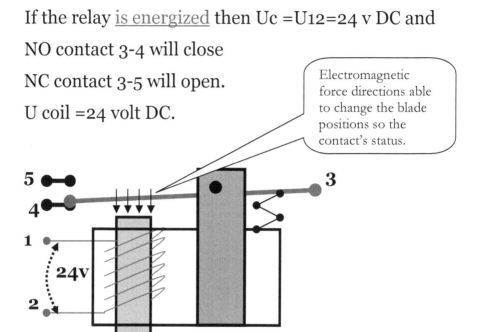

Diagram of contacts:

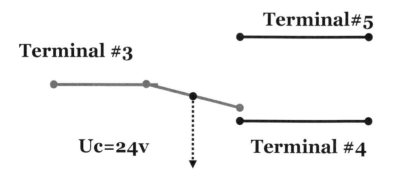

The COIL is the main part of any relay and will be rated at either: 12 volts, 24 volts, 48 volts, 120 volts or 240 volts. The relay's coil will be energized with AC or DC voltage as specified before.

To summarize, when a voltage is applied to the coil at terminals 1&2, the status of the relay's contact will change. We've examined the way in which a relay works. This means having the coil de-energized or energized. Let's now explore this relay within a control diagram in order to observe its functions.

Let take a previous example on page 24 where a logic diagram was required to control the air pressure for an air tank. In the dotted area we need to integrate our relay and contacts to realize the function of the logic diagram. Thus we need to "build" the logic diagram within the relay and use its contacts as integrated parts of the compressor's control diagram. The next series of diagrams will utilize both DC and AC relays (see the power source P/S).

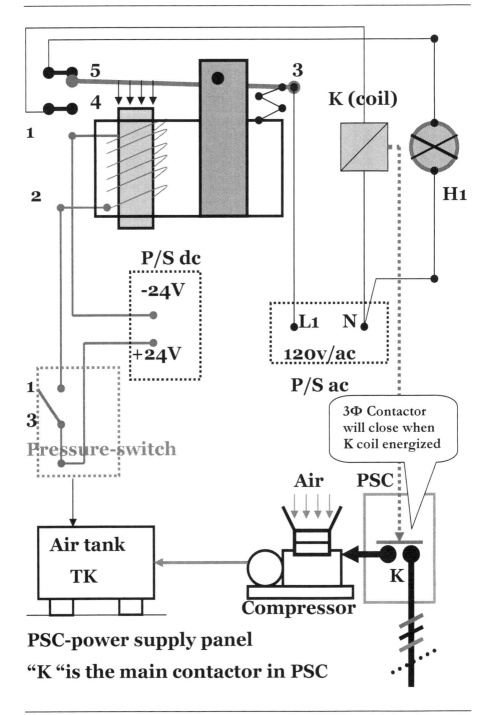

PSC-power supply panel

"K "is the main contactor in PSC

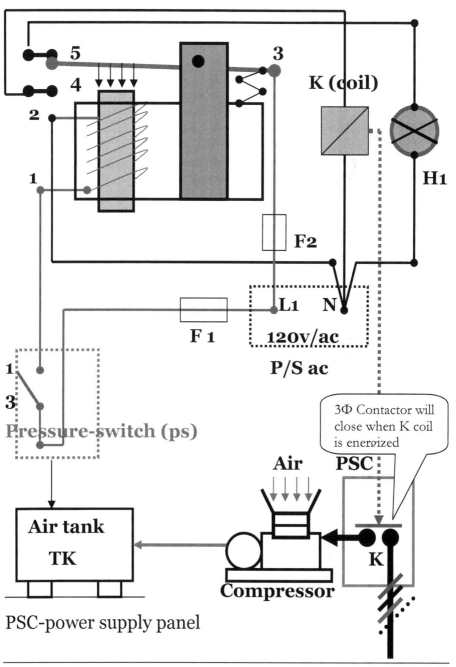

K (coil)

H1

F2

F 1   L1   N

120v/ac

P/S ac

1

3

Pressure-switch (ps)

3Φ Contactor will close when K coil is energized

Air   PSC

Air tank

TK

Compressor

K

PSC-power supply panel

"K "is the main contactor in the Power Supply Panel.
The diagram illustrates two options:

1   Relay's coil at 120 volt AC(page 43)

2   Relay's coil at 24 volt DC (page 45)

A logic diagram using the appropriate symbols:

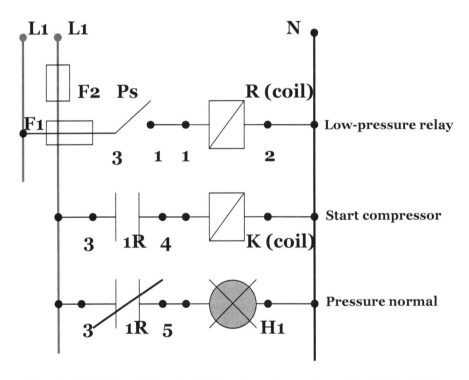

A logic diagram will display the contacts' status when they are on the shelf storage (in other words, when the device they belong to is not energized or activated).

As a result of being energized or activated, the device will change the status of its contacts. This activated state is the opposite of what is represented in the diagram.

The next illustration is a logic diagram for a relay coil at 24 volts DC. Again, the proper symbols are used.

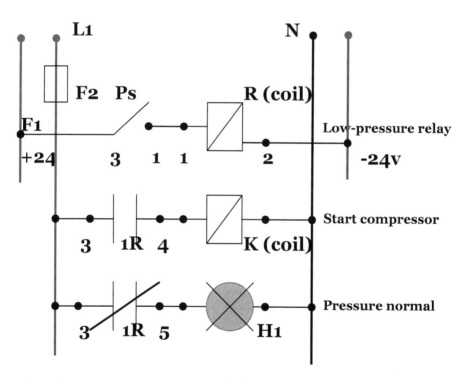

The functions of the relay in both the logic diagram on page 42 and that on page 43 are the same. In fact, the only difference between the two is that the power supply for each relay differs. On page 45 the relay is working in DC(direct current) and in the illustration on page 44 the relay is working in AC(alternative current). This means that a relay device can work at different voltage type (AC or DC) or levels (24v or120v) yet provide the same function.

Any control diagram may contain more than one relay and pair of contacts. The more conditions included in a logic diagram, the more relays a control diagram will have.

CONTACTORS

Contactors are almost identical to relays. They work on the same concept and are composed of a coil, magnetic core and contacts. The difference between a contactor and relay is that contactors are responsible for connecting and disconnecting the power within a power circuit, so it is designed to close or open currents higher than those going through control devices. As an electrician, you will discover that almost all applications contain this device. Again, the contactor is specifically designed to connect and disconnect the power for electrical equipment. Contactors are available in single phase (one pole/1P), two phase (two poles/2P) or three phase (three poles/3P) options for AC distribution and DC distribution. They are rated based on the following factors:

Voltage (AC or DC) and Amperage

A contactor's electro-mechanics:

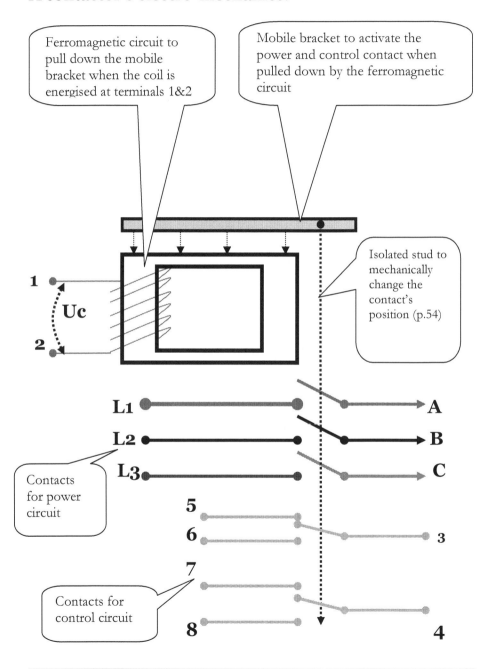

Ferromagnetic circuit to pull down the mobile bracket when the coil is energised at terminals 1&2

Mobile bracket to activate the power and control contact when pulled down by the ferromagnetic circuit

1

Uc

2

Isolated stud to mechanically change the contact's position (p.54)

L1 ●━━━━━━● ╲ ● ➔ **A**

L2 ●━━━━━━● ● ➔ **B**

L3 ●━━━━━━● ● ➔ **C**

Contacts for power circuit

5 ●━━━━●
6 ●━━━━● ━━● **3**

7

Contacts for control circuit

8 ●━━━━● **4**

Upon examination of the previous diagram, we can observe on the contactor that there are two sets of contacts designated for control: 3-5-6 and 4-7-8.

You will discover contactors with or without control contacts. Most of the contactors that have power contacts available will only provide spaces for control contacts to be attached to it. This is a quality feature since it makes possible the attachment of auxiliary contacts when required. This also allows the control diagram to be more flexible.

Power contacts will be able to carry a variation in current. For example, a current ranging from 6 amps to 200 amps may be carried by a contactor. L1/L2/L3 is incoming side for the current while the A/B/C is the load side (see diagram on page 48).

A control contact is also able to carry a variation in current. For instance, a control contact may carry a current which varies from 1 amp to 10 amps.

Contactors are enclosed in electrical boxes or open devices inside the electrical panels. The control circuit energizes contactors. For special applications you may find contactors that seem to be much more complicated but in fact the working concept is the same.

In some applications you may also find something called a "dual voltage coil." This simply means that the coil will be supplied at either 120 volts OR 240 volts. Some people are confused by the term dual voltage and interpret it as meaning that the coil requires BOTH 120 volt AND 240 volt power.

As you've likely already discovered, this book specifically deals with control circuits. Therefore, we are only touching briefly on other subjects that contribute to our understanding of control circuits.

The supplementary books in my series will fully address issues and subjects that are critical for an electrician to understand. The series will provide a complete "electrician's library."

Each book will contain essential and detailed information on specific subjects.

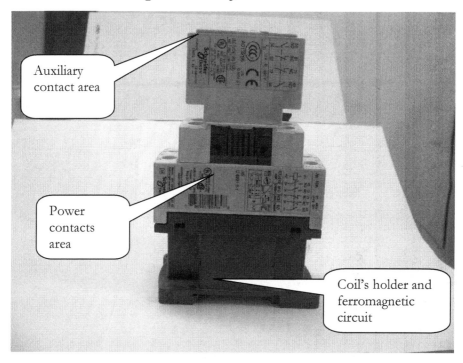

Auxiliary contact area

Power contacts area

Coil's holder and ferromagnetic circuit

This is a contactor , your notes here:

CONTACTOR SCHEMATICS:

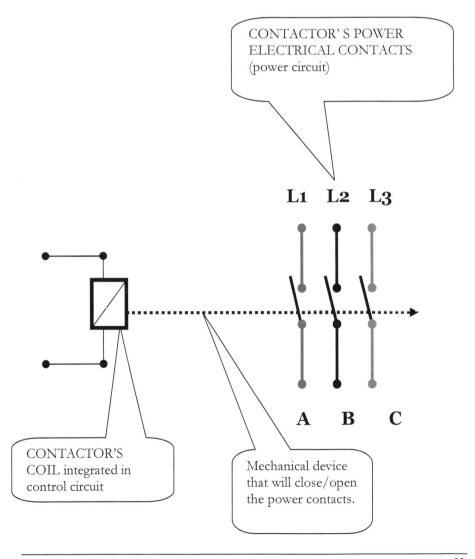

CONTACTOR' S POWER
ELECTRICAL CONTACTS
(power circuit)

L1 L2 L3

A   B   C

CONTACTOR'S
COIL integrated in
control circuit

Mechanical device
that will close/open
the power contacts.

Coil's holder and ferromagnetic circuit (bottom section)

Coil's terminals

CONTACTOR'S COIL integrated in control circuit

Mobile bracket to engage the power and controls contact when pulled down by the ferromagnetic circuit

Spring to hold the mobile ferromagnetic section away from the bottom section of the ferromagnetic circuit.

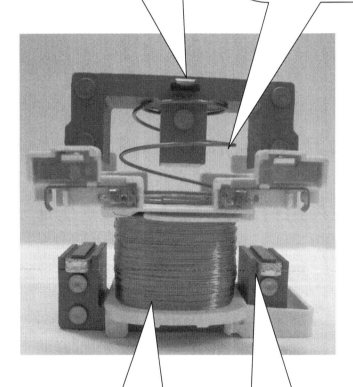

Coil and ferromagnetic circuit (bottom section)

Ferromagnetic circuit (bottom section)

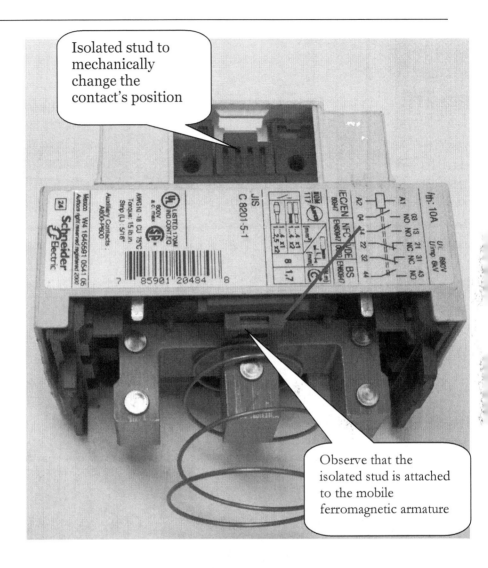

Isolated stud to mechanically change the contact's position

Observe that the isolated stud is attached to the mobile ferromagnetic armature

Observe the "gap" between these two top and bottom sections of the ferromagnetic circuit when the coil is **not energized.**

There is no "gap' when the coil **is energized**. The top section (mobile) of the ferromagnetic circuit is attracted by the bottom section. The spring action is cancelled by the electromagnetic force and the "gap" is not there anymore. The top section is being attached to the power contacts via the blue stud (see page 54) will change contact's status to "close". Since the coil will be de-energized the electromagnetic force will be zero (Fe =0) so the spring will create the "gap "in between the top and bottom sections of the ferromagnetic circuit so the power contact will "open".

This picture is displaying the position of the top section of the ferromagnetic circuit. Here we've created a simple model by pressing the top section with the finger **to simulate** the action of the electromagnetic force.

DON'T TOUCH ALIVE ELECTRICAL DEVICES

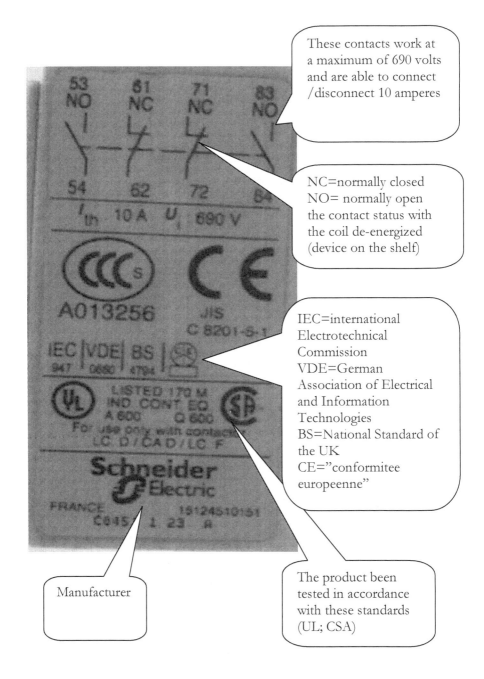

These contacts work at a maximum of 690 volts and are able to connect /disconnect 10 amperes

NC=normally closed
NO= normally open
the contact status with the coil de-energized (device on the shelf)

IEC=international Electrotechnical Commission
VDE=German Association of Electrical and Information Technologies
BS=National Standard of the UK
CE="conformitee europeenne"

Manufacturer

The product been tested in accordance with these standards (UL; CSA)

You must read the labels on your devices! They provide a wealth of information.

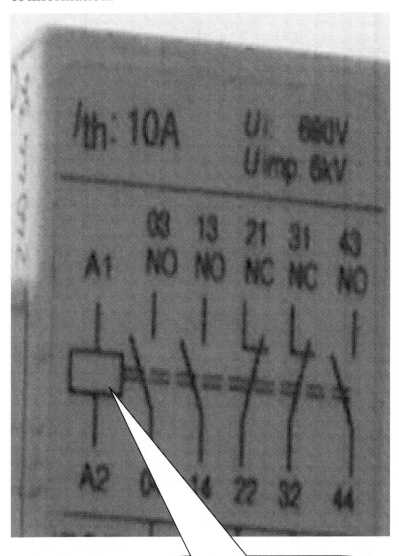

Please observe this symbol as the coil of the contactor and the coil's terminals A1 and A2

Auxiliary control contacts.

Here are the contacts to terminate the power

NO or NC

In my experience, many skilled electricians have a hard time understanding the contact configuration belonging to a contactor, relay or other device, such as a pressure switch. Due to this fact, I will share my experience to ensure you fully understand this concept. This will guarantee you don't make any significant errors, such as confusing NO(NORMALLY OPEN) and NC (NORMALLY CLOSE)in a contact configuration. It is important you understand when a contact is NO or NC. THIS is NO (normally open) contact:

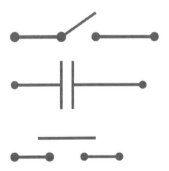

In this representation we have symbols that are to be used to indicate NO contact. The number of the NO or NC contacts for a relay or contactor are the numbers of contacts we can find OPEN or CLOSE having the relay or contactor stored on the shelf in the warehouse.(this being de-energized status)

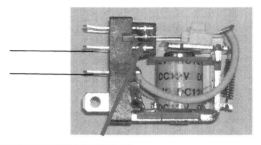

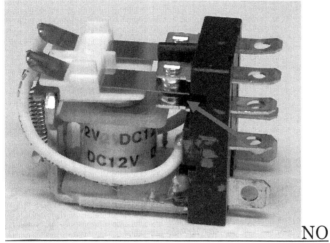

NO

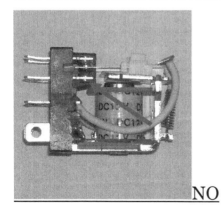

NO

When the relay is energized the status will change and the NO (normally open) contact will CLOSE.

THIS is NC (normally closed) contact:

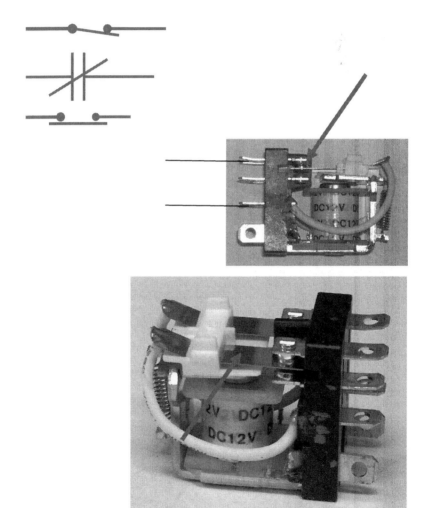

In this representation we have symbols that are to be used to indicate NC contact.

The number of the NC contacts for a relay or contactor are the numbers of contacts we can find CLOSE having the relay or contactor stored on the shelf in the warehouse (this being de-energized status)

NC: normally closed contact

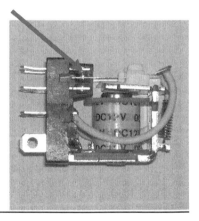

When the relay is energized the status will change and the NC (normally closed) contact will OPEN.

As you can see there is no power applied to the relay's coil.

WHY WE NEED CONTROL CIRCUITS

In all applications, control circuits are designed <u>to control</u> (NOT TO CONNECT) the power supply for devices such as: electrical motors, heaters, electrical pumps, electrical equipment, transformers etc. Devices such electrical motors, heaters, electrical pumps, equipment and transformers will include connection devices located on their power circuit to connect or disconnect the power for them. The bellow diagram will explain this:

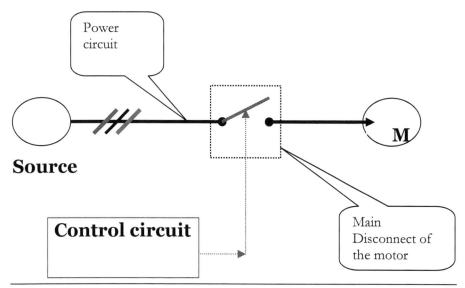

These control circuits will include wires, relays, contacts, sensors and automatic switches. They are connected to a logic diagram.

In other words, these circuits will consider the status/conditions of the parameters into the logic diagram and will generate as decision the connection or disconnection of the power supply for specific devices. Control circuits CONTROL these devices. Hence the name: control circuit.

Within the industry there are different standards but most of the time control circuits operate at a very low voltage, usually less than 120/110 volts. They are provided with 6v, 12v, 24v or 48v in either DC or AC currents. Control circuits have extensive applications and, therefore, a solid knowledge and understanding of them is a prerequisite for electricians to install, repair and maintain control circuits for complex devices or installations. Modern control circuits are used in cutting-edge technology and are often responsible for reducing the amount of wiring and connections between elements of the logic diagram. Sophisticated devices are computerized and programmable and need little maintenance as they are more accurate and very quick in response times. Devices such as these also require less space and yet provide high-quality performance. As an electrician, if you have a solid grasp of the way in which a control circuit works within any device, you'll be able to apply this knowledge to understand more sophisticated applications.

As an electrician, it is essential that you are have a solid knowledge and are comfortable with the basic workings of a control circuit since you will be required to be a part of the construction and/or maintenance team as a part of your daily activities. Most control circuits are an element of the logic diagram and can be read and understood within that process.

By trying to understand the technological process of the control circuit, you'll begin to understand the logic diagram. As designs for electrical equipment have become more sophisticated, in turn control diagrams have become more complicated. As part of this evolution, programmable logic controllers (PLC) have been created. These are also known as smart devices. In order to maintain and program these devices, you will have to have a firm knowledge in the basics of control circuitry.

Working in the PLC/smart devices field is very challenging. Therefore, in my opinion you should start your career working with the basics first. I've seen many individuals enter the PLC field hoping to immediately find a high paying job. It is essential that an electrician has a good knowledge of the basics of the industry before entering more complex areas of expertise.

My advice is simple. Once you have gained experience working with traditional control diagrams, then you can move on to more complicated devices such as programmable logic controllers (PLCs).

Some final words about programmable logic controllers (PLC): They were introduced into the industry in 1969; about the same time that electronic devices were replacing hardwired devices. With the computer revolution, the role of programmable logic controllers in the industry has steadily increased, specifically in the fabrication process. The first PLC was invented by Richard E. Morley in 1969. He was the founder of the Modicon Corporation. Over time this electronic device has become more user-friendly. Innovations and upgrades have made it feasible to work with the device even if one has little knowledge of electronics.

POWER CIRCUITS

In most applications, power circuits are either single-phase circuits or three-phase circuits.

Power circuits are designed <u>to supply power</u> to electrical devices such as: motors, heaters, electrical pumps, equipments, transformers, etc. The diagram below illustrates a three-phase power circuit (single line diagram).

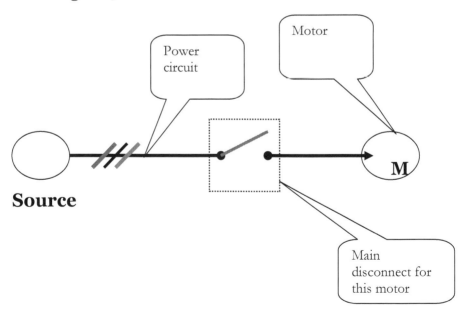

These circuits include cables, splitters, and contactors and fusible or non-fusible disconnects. They supply power for the loads at the time and conditions established by the logic diagrams through the control circuits.

It easy to remember why these are named power circuits since they are the power/electricity supply paths for the loads.

WHAT IS THE DIFERENCE BETWEEN CONTROL AND POWER CIRCUITS?

There are essential differences between these two kinds of circuits:

1. Control circuits represent the "brain" of the equipment and, by way of the logic diagram, decide to connect or disconnect the power supply for a specific load.

2. Power circuits connect and disconnect the power supply at the load when required by the control circuit logic diagram. Power circuits are the physical support that transport and carry energy thorough lines to the load. Power circuits are designed to supply power to the loads within safe conditions.

3. Higher currents travel along power circuits and lower currents on control circuits.

HOW TO READ CONTROL DIAGRAMS

We will start with an easy control diagram as been presented chapters before at page 43and the next one at page 74. The status of the next control diagram is indicated by the green indication light (H2). When the air pressure in the air tank is in normal range, the green indication light will be on(H2). The H4 indicating light is ON and shows the compressor status "OFF" which makes sense since as the air pressure is in normal range. The device that will trigger the diagram to switch into another status is called a **pressure switch.** By examining the diagram, we can see this device is Ps (Pressure Switch) and its contacts Ps 1-3. The **H1 light indicator** for "tank Pressure-low" is OFF. Relay R is not energized. The contactor's coil K is also not energized, so the compressor is OFF. On the contactor there is a pair of control contacts 1K and 2K that are integrated into lines "d" and "e" which are the indicating lines for the compressor's status. These symbols make the diagram easier to read. All the devices in this diagram are working at 120-volt alternative current, which is the voltage level for the control diagram. Remember we may have 24 volts or 120 volts and power in AC or DC. Safety requirements and/or environmental conditions will often determine the voltage level.

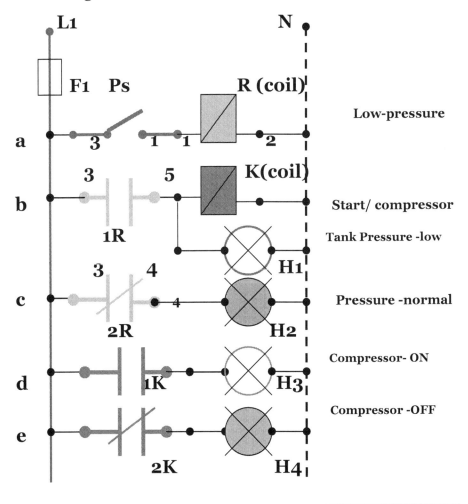

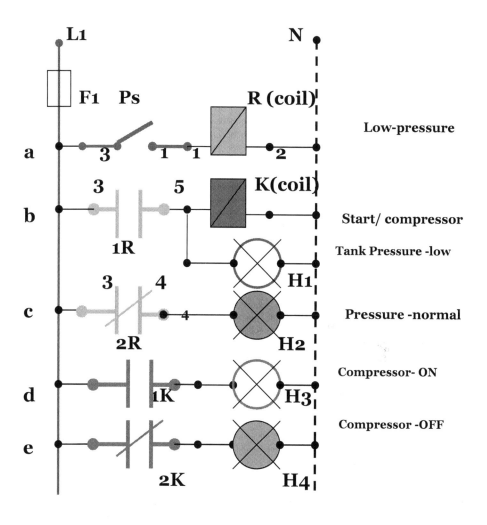

L1 is representing the HOT line, N is representing the Neutral line so it is Black/**White**

SYMBOLS

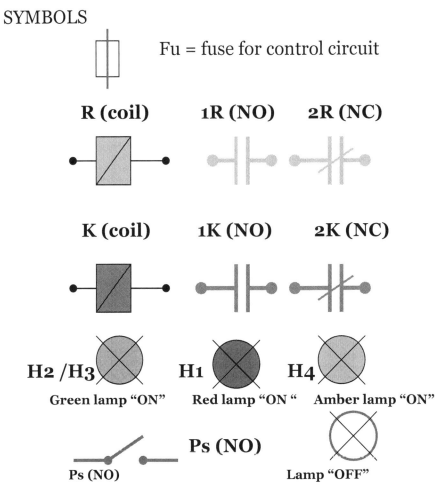

Fu = fuse for control circuit

R (coil)   **1R (NO)**   **2R (NC)**

K (coil)   **1K (NO)**   **2K (NC)**

H2 /H3 Green lamp "ON"   **H1** Red lamp "ON "   **H4** Amber lamp "ON"

Ps (NO)

Ps (NO)   Lamp "OFF"

For easy understanding **the coil and contacts belonging to the <u>same relay</u> will have the same color** in our diagrams. In this diagram there is: one relay(R), one-contactor (K), four indicating lamps (H1; H2; H3; H4) and a pressure switch (Ps).We are showing the pressure switch (which initiates all the tasks) as ACTIVATED.

The **H1** lamp acts as an alarm indicating that there is **air low-pressure inside the tank.**

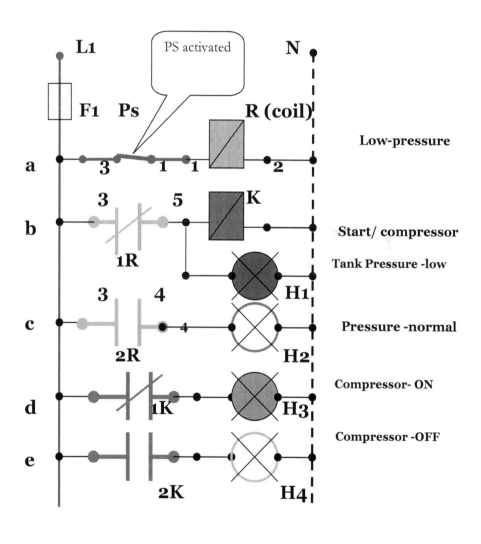

Please note that the diagram is triggered by the NO contact of the pressure switch PS (1-3). When this closes at PS1-3, it will energize the R coil and the diagram will switch from the activation stage shown in page 73 to the stage shown in page 77 and/or page 79.

1. At this point you need to examine all contacts belonging to "R" relay. As you will observe in the logic diagram on page 72, this relay has two contacts 1R (NO at 3-5) and 2R (NC at 3-4).

2. 1R will close and as a result will energize the contactor's coil K and the lamp H1 (indicating low pressure in the air tank). Contactor K will then close its power contacts and will energize the compressor. The compressor will then supply the tank with more air.

3. 2R will open and as a result will de-energize the green lamp H2 (pressure normal). It will stay off since there is no "Pressure –normal"

4. The contactor will close its control contacts 1K and will then supply the H3, thus confirming

that the compressor is running (compressor ON).The contactor will open its control contact 2K and as result will cut the power supply for H4 (shutting the compressor off). The diagram will change in this way:

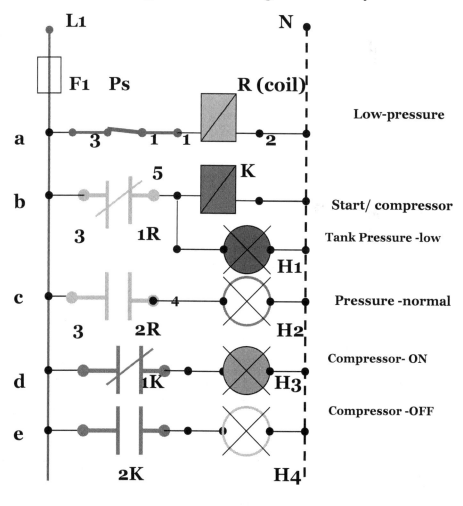

<u>Thus the status of each device from lines "a" to "e" has been changed and the compressor is running which will supply the air tank.</u>

Throughout the time the compressor is running, the pressure switch will have the contact Ps 1-3 closed. By referring to the diagram on page 79, we can see that this will allow the equipment (compressor) to be energized by contactor K until the air pressure inside of the tank is within the prescribed limits (see page 42&43). This process is controlled by the pressure switch PS contact 1-3 which ensures each step is taken until the diagram reaches the status indicated on the next page. In our example we envisioned the diagram as having all the control components (relay, contactor, lamps and pressure switch) operating at a 120 volt alternative current. The control voltage is provided by an alternative power source at 120 volts. Any short-circuit on the control side will be eliminated by the F1 fuse. So "F1" will act as a protective element for our diagram. Thus the protective element is a fuse. The diagram will change to its original status as displayed in the following diagram. When the pressure reaches the prescribed limits, the contact Ps1-3 will open and the diagram will exhibit the initial status: Pressure-normal /compressor off

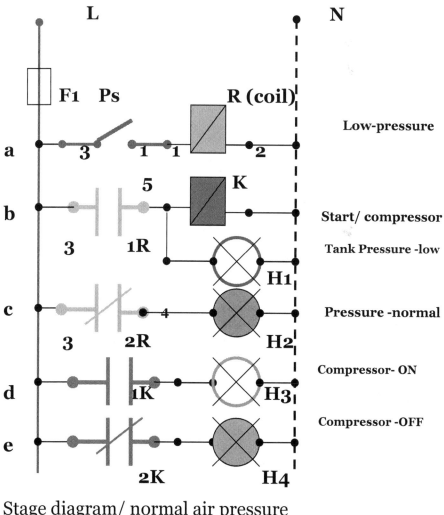

Stage diagram/ normal air pressure

Again to repeat the sequence of this diagram for better understanding:

When the pressure reaches the prescribed limits, the contact Ps1-3 will open and the diagram will exhibit the initial status:

1. Relay R de-energized
2. Contact 1R NO will open
3. Contactor's coil K will be de-energized
4. H1 goes OFF (low pressure)
5. Compressor will stop
6. Contact 2R NC will close
7. H2 is ON −pressure normal
8. 1K /NO is open H3 goes OFF
9. 2K /NC is closed amber light H4 is ON (compressor OFF-stops)

EXAMPLES OF CONTROL DIAGRAM

We are now going to a more detailed examination some of the control diagrams you will most often come across during your career as electrician. These control diagrams are:

1. Exterior lighting control by sensor/timer

2. Motor spin direction change

3. Exhaust fan $1\Phi/3\Phi$, control diagram

4. Magnetic starter control diagram

5. 3-way switch lighting control

6. 4-way switch lighting control

Over the next few pages we will analyze each of these diagrams individually. My hope is that you will have an opportunity to exercise this knowledge and practice in a hands-on way in order to really take advantage of the lessons in basic control circuitry covered within this book.

More sophisticated diagrams such as control diagrams for elevators, cranes and other complicated machinery will require further, specialized training but this book definitely gives you a foundation in the basics of control circuitry. As an electrician or electrician's apprentice, you can think of it as your "first step" in learning about control circuit.

Most of you will not have the opportunity to work with these kinds of installations during your first years on the job.

However, you should continue reading and learning as it this will ensure that you are well-informed. As well, it will build your ability to learn fast and enable you with a basic understanding of some more complicated tasks.

Then, when an experienced electrician or foreman will asks you to work on one of these tasks or installations, you will feel more confident tackling the job!

EXTERIOR LIGHTING SYSTEM CONTROLLED BY LIGHTING SENSOR

When the control is based on the sunlight presence, this will require the installation of a photocell device in an area where no obstacles will block its access to sunlight. This means that most of the time this type of system is located on the southeast wall of a building where the lighting is controlled.

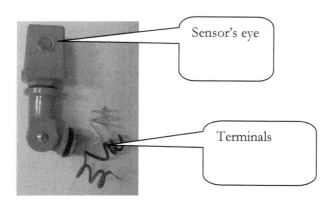

Sensor's eye

Terminals

The sunlight will activate the sensor and keep the light switch located on the sensor OPEN so the lighting system will be OFF during the day. When the sun sets and the quantity of light decreases, the sensor will switch its contact to the CLOSED position. The lighting system will then be ON during the night. This sensor will control a contactor that will connect and disconnect the power for the entire exterior lighting system.

Box connector and terminals

Sensor's "head"

Here is the diagram showing the sensor's status during the day:

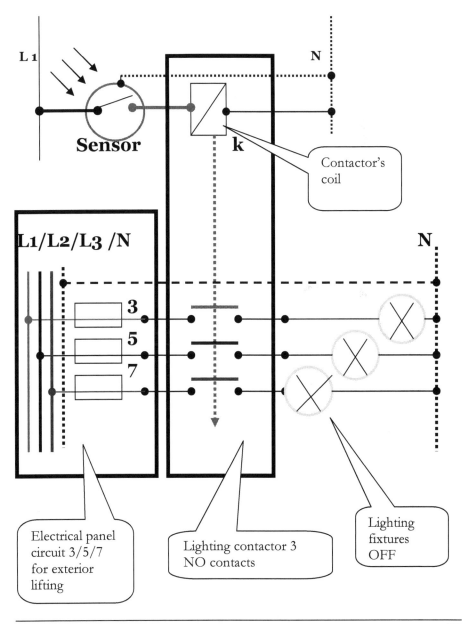

L 1

N

Sensor

k

Contactor's coil

L1/L2/L3 /N

N

3

5

7

Electrical panel circuit 3/5/7 for exterior lifting

Lighting contactor 3 NO contacts

Lighting fixtures OFF

L1/L2/L3 =120/120/120 or 347/347/347 volt

Night sensor status:

U=120v/ac

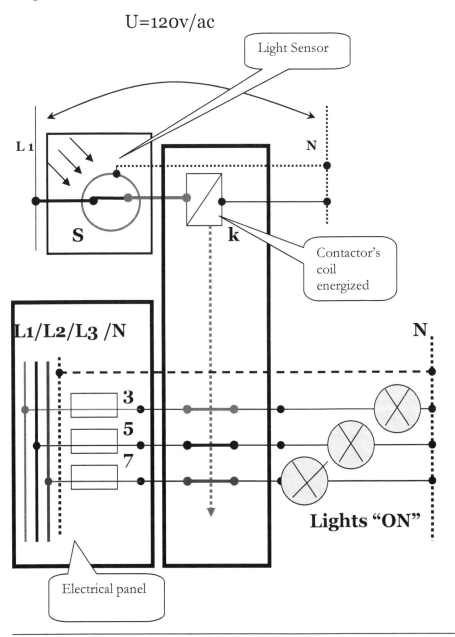

Most of the photoelectric sensors will require a 3-wire connection as shown in pages 87&88. These wires are usually red, black and white. This device needs power at hot (black) and neutral (white) in order to work. One wire is the output from the sensor when it is activated. This output will supply the lighting fixture.

1. During nights the sensor contact that is NO (normally open) will close.

2. The contactor's coil will be supplied at 120 volts AC and will close the power contacts supplying the lighting fixtures located on the circuit 3/5/7, enabling it with power so the lights will become ON (yellow status)

Note: Observe the protection devices for each circuit (3/5/7). These protection devices are fuses or circuit breakers. Any short circuit that occurs will trigger the protection device to work. The protection device will then disconnect power for the affected circuit.

Fuses can get melted and circuit breakers will trip and can disconnect the power for the load of that branch circuit (3/5/7).

EXTERIOR LIGHTING SYSTEM CONTROLLED BY TIMER DEVICE

When the control is required to be created by a timer device, we need to install a timer device in an area of the electrical panel that will allow one contactor to control the lighting system. These timers are electromechanical or the electronic type. They will display the day's hours and the night's hours and the current hour as well. The main part of the timer is **the synchronous motor** which works at 120-volt AC (See pages 97 or 98). This motor is accurate and will not create errors as these timers are very exact in their timing. These devices will also have 1 set of pins (or trippers- see page 94) that will switch the power to the loads ON or OFF as per programming.

At the front of the device there is a metallic (yellow) disc with the 24 hours of the day and night printed on it. On the front of the disc you should locate the pins (trippers). For the day sequence you need the power for lighting to be cut off. You also need to be able to locate the pin on the front of the disc for the <u>night sequence.</u> This is important when you need the power to energize the lighting system.

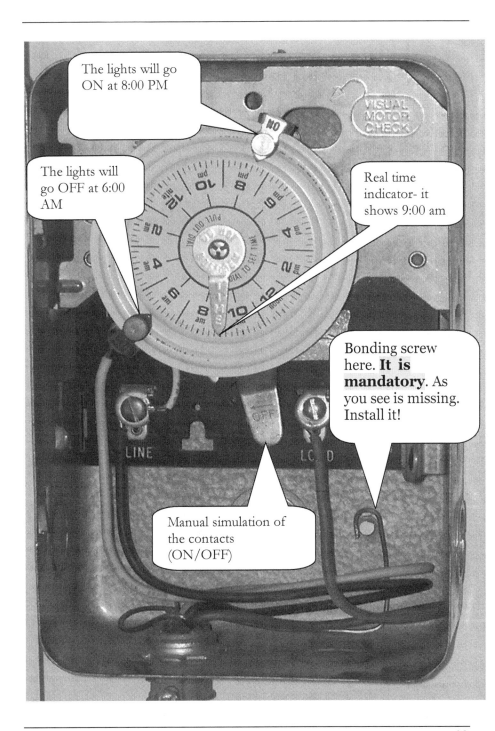

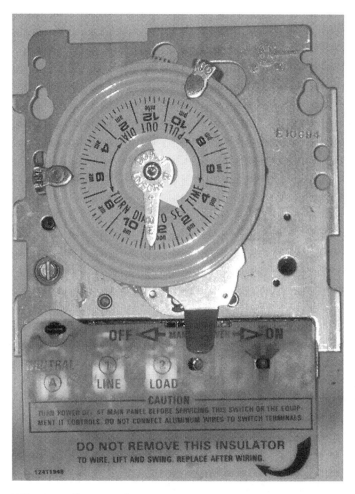

Timer- front view

Observe the insulator protects the terminals from directly touching. They may be alive!!!.

Trippers or pins: ON(left) and OFF(right). Observe the screw that holds trippers on the yellow disc.

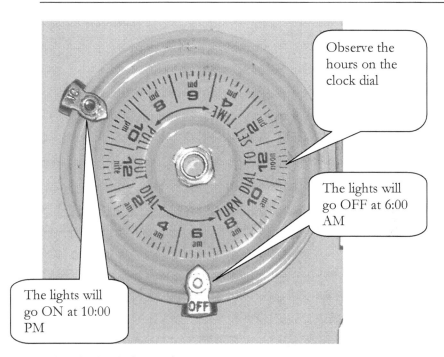

Clock dial disc; front view

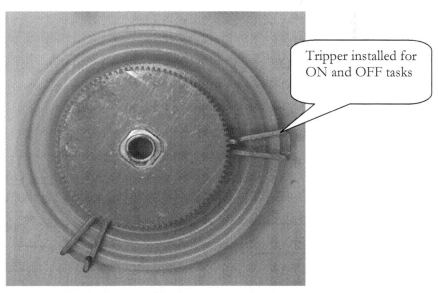

Clock dial disc; back view

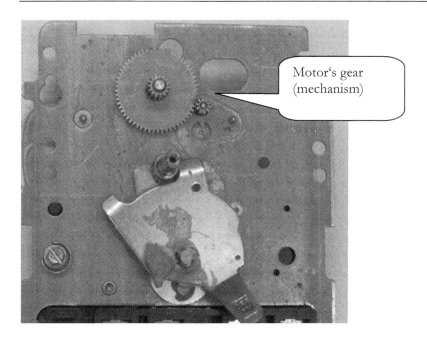

The mechanics can be observed very well in this picture, helping you understand the way in which this device is working.

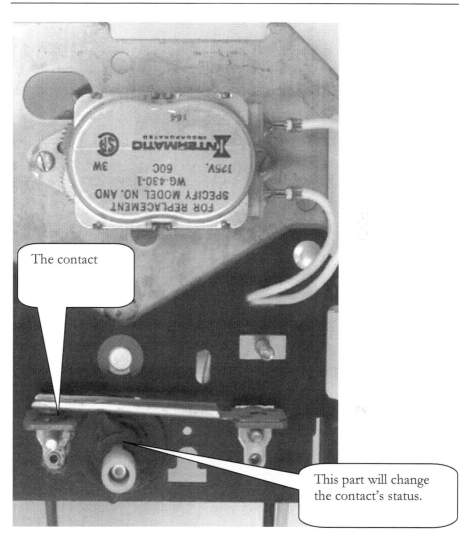

Back view of device:

Observe the motor, the contact and the back plate.

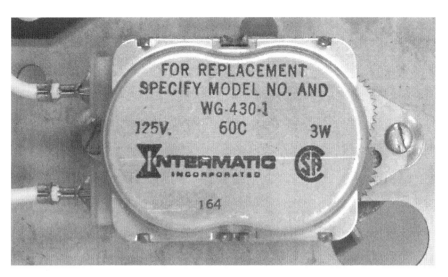

This is the motor with its characteristics and model #

The contact will cut or connect the power for the load side.

PROGRAMMING INSTRUCTIONS

1. **TO SET "ON" AND "OFF" TIMES:** Hold trippers against edge of **CLOCK-DIAL**, pointing to time (AM or PM) when **ON** and **OFF** operations are desired, tighten tripper screws firmly. For additional tripper pairs on **CLOCK-DIAL** order 156T1978A.

2. **TO SET TIME-OF-DAY:** Pull **CLOCK-DIAL** outward. Turn in either direction and align the exact time-of-day on the **CLOCK-DIAL** (the time now, when switch is being put into operation) to the pointer. **DO NOT MOVE POINTER.**

OPERATING INSTRUCTIONS

- **TO OPERATE SWITCH MANUALLY:** Move **MANUAL LEVER** below **CLOCK-DIAL** left or right as indicated by arrows. This will not effect next operation.

- **IN CASE OF POWER FAILURE**, reset **CLOCK-DIAL** to proper time-of-day. See programming instructions.

INTERMATIC INCORPORATED
SPRING GROVE, ILLINOIS 60081-9698

1587S1094:

The manufacturer in this case is INTERMATIC. To get information on the operation of devices such as this, the internet is powerful tool. Don't hesitate to search online for information.

The next few diagrams will use representations of the timer's parts to help you understand the way in which this device works.

Diagram of timer

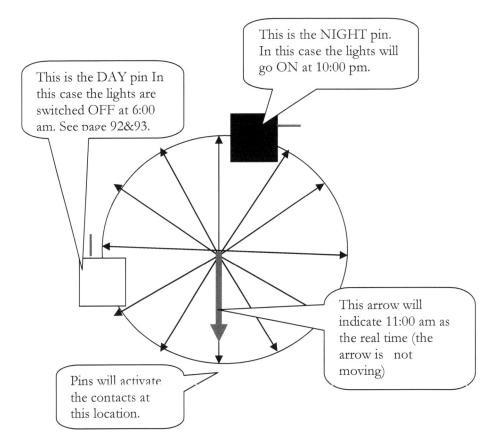

When the disc is moving, the arrow will not move. After you've pointed the trippers at the time indicated to connect or disconnect the lighting system, pull the disc forward and rotate it clockwise. You'll find the real

hour or time is indicted by the arrow. In our case the
hour shows 10:00 a.m.

The main parts of the TIMER:

(This timer is set for a few minutes before 10:00 p.m.
when the lights will turn ON)

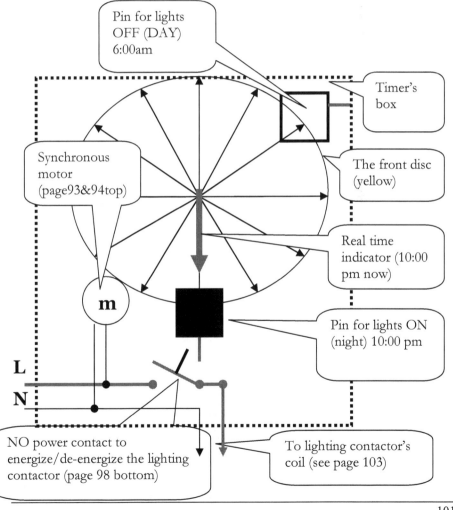

Pin for lights OFF (DAY) 6:00am

Timer's box

Synchronous motor (page93&94top)

The front disc (yellow)

Real time indicator (10:00 pm now)

m

Pin for lights ON (night) 10:00 pm

L

N

NO power contact to energize/de-energize the lighting contactor (page 98 bottom)

To lighting contactor's coil (see page 103)

When the timer has the power supply "ON," the synchronous motor will activate the mechanism and start running. The front disc (yellow disc) will turn around clockwise and the pins located at the set hour for connecting the power will reach the timer's contact (page 97&98). This will change the status from open to closed. It also means that power will be provided to the contactor's coil in much the same way as shown earlier. A lighting contactor will switch ON all the lights located on the circuits allocated for exterior lighting (page 103). A light goes "ON" at this point. The next page will explain in detail this sequence by showing a control diagram a few minutes before and after 10:00pm. This will provide a clearer explanation of the way in which the timer energizes the lighting fixtures. The first diagram shows the arrow is indicating that it is nearly 10:00 pm. The lights will go ON at 10:00 p.m. At that time the black pin will closes the contact (see page 97/98). This will energize the contactor's coil and trigger the diagram to change from "lighting OFF" status to the stage of "lights ON" as shown on page

101.A few minutes <u>before</u> 10:00 Post Meridian (P.M.)

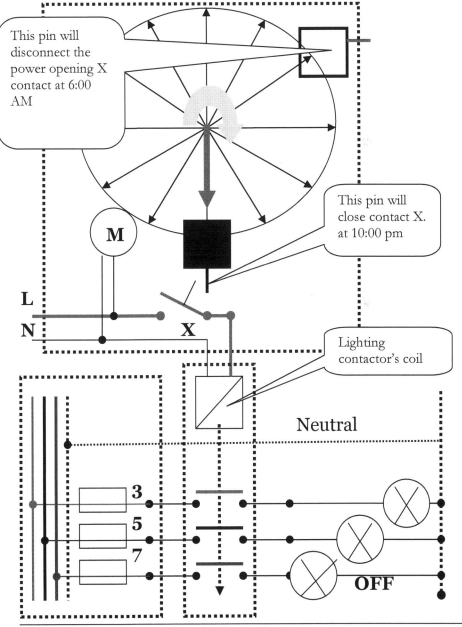

This pin will disconnect the power opening X contact at 6:00 AM

This pin will close contact X. at 10:00 pm

Lighting contactor's coil

M

L

N

X

Neutral

3

5

7

OFF

An hour after 10:00 Post Meridian (P.M.)

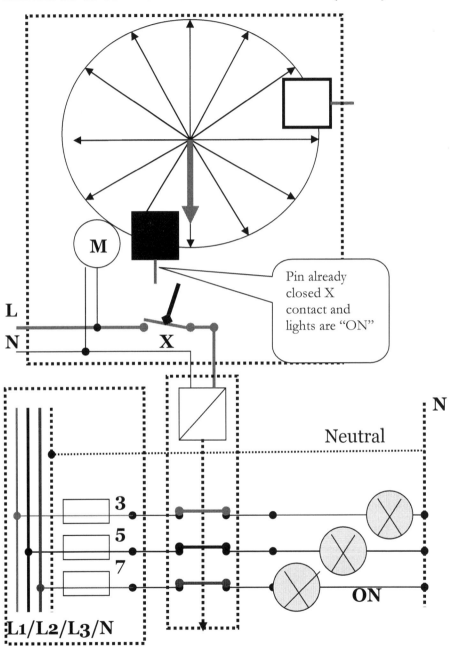

Pin already closed X contact and lights are "ON"

SPIN DIRECTION CHANGE FOR 3 Φ MOTOR
(Control diagram)

In order to change the rotation for a 3 phase motor we need to switch the supply line's phases. Here we will assume that the sequence of the power supply is L1/L2/L3, with a clockwise rotation (rotary motion). By changing the power supply sequence to L2/L1/L3, the motor's rotary motion will change from clockwise to counterclockwise (the opposite).

 We can do this by disconnecting the motor and changing the terminals.

This will change the rotation. Some motors are required to change rotation in order to perform some specific tasks.

There are control diagrams able to do this automatically when required.

You may be aware of other ways to do this but most of the time you'll find this method is applicable.

Since we are talking about the change of power supply to the motor, we need to have two contactors that connect the power on the power circuit to the motor in different sequence but not simultaneously.

One contactor will supply power in sequence L1/L2/L3 since other in the different sequence for example L1/L3/L2.

These will allow the motor to be energized by the power supply line in different sequences.

In order to prevent these two contactors from switching the power ON at the same time to the motor, they are interlocked mechanically or electrically.

Interlocking means will not be possible way to connect them simultaneously to provide with power to the motor.

The block diagram for the power circuit and control logic would look like this:

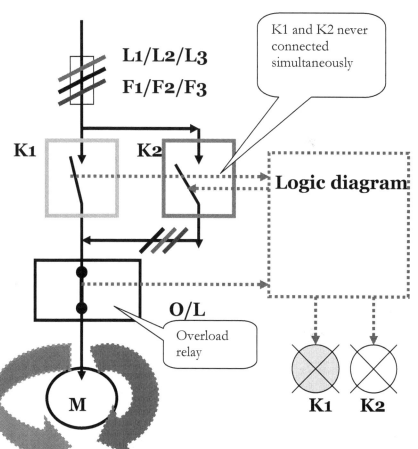

As you can see the diagram is showing the line, the contactors, the logic diagram and the overload relay. Also the two blue arrows will show the left and right spin direction. Two indicating lamps will provide info about status of the electrical diagram.

The following image will help you understand the connection diagram:

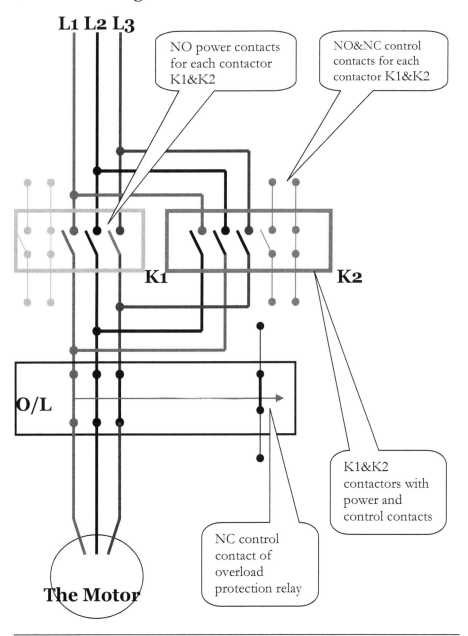

The illustration shown on page 107&108 is the power circuit diagram for a 3-phase motor. It displays the main parts of the power circuit. It also indicates the control contacts that are to be used in the logic diagram. These will determine the rotational change when such a change is required by the operator or process. The overload relay symbol O/L will work any time an overload occurs at the motor.

An overload might occur as a result of following reasons:

1. Mechanical overload in the equipment powered by the electrical motor. Most of the time this is the reason for overload.

2. Wire defects within the motor's coils. In this situation, the motor is already compromised but the relay will protect the installation from developing major defects or equipment damage.

The overload relay is designed to work in case of an increase in power that exceeds the nominal current by 10% to 15 %. It is obvious that any overload will increase the nominal current, so based on this information the O/L relay will work and trigger a task to disconnect the power supply of the motor. After elimination of the overload, the motor can be connected to the power supply again. Most overload relays will be provided with a "reset push button". The reset of the O/L contact is required since the contact will not return to the NO position after the overload is eliminated (See pages: 124&125, 126 and 127).

In next diagram I will explain how the overload relay works as a result of an overload. The overload relay is the device having a current path with three plates, one for each phase and installed on an isolated support. In the composition of each plate there are two materials with different characteristics of dilatations. When an overload current passes each path, it will generate dilatations on the overloaded phase. The result will be the bending of the plates which will activate a mechanical device (PVC bracket) to open a contact located in the control diagram. This contact will generate the disconnecting task through the control diagram protecting the motor.

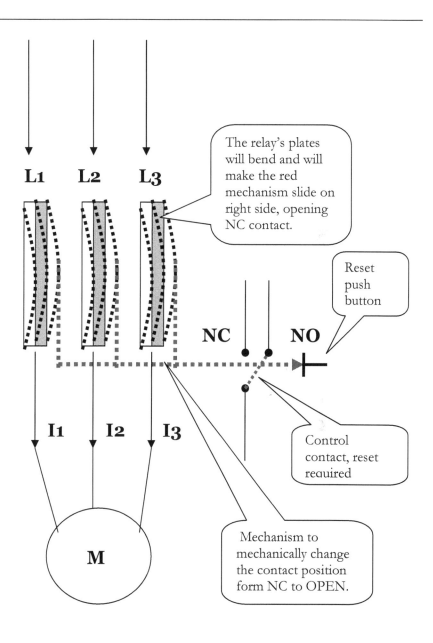

Why will the plates bend? Having materials with different characteristics of dilatations attached together will make the plates bend. This will make the mechanism to slide right (page 111). Thus reset is required to reposition the contact in NC position. Below is a diagram to further this explanation:

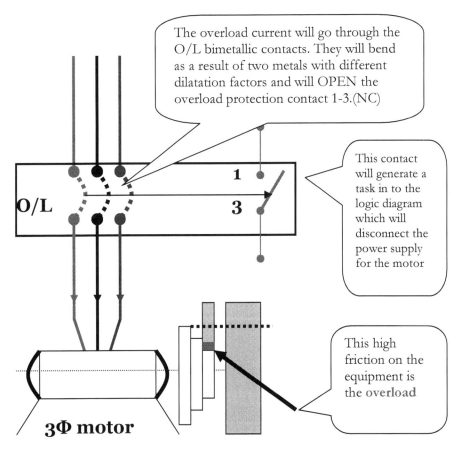

Equipment powered by this motor

Now since we understand the O/L relay lets examine the control circuit and power circuit symbols.

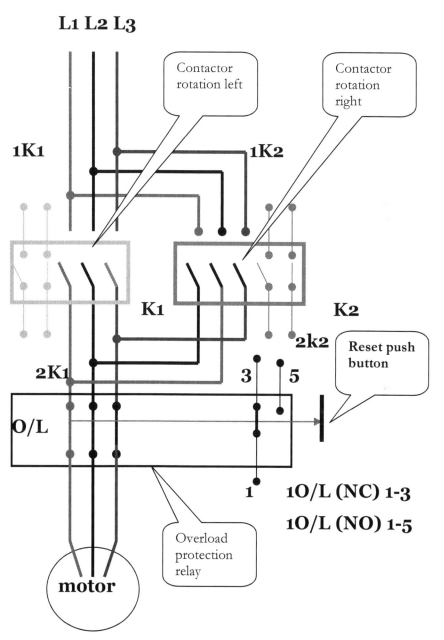

Associated with this single line diagram will be a control diagram as presented next page:

Control diagram

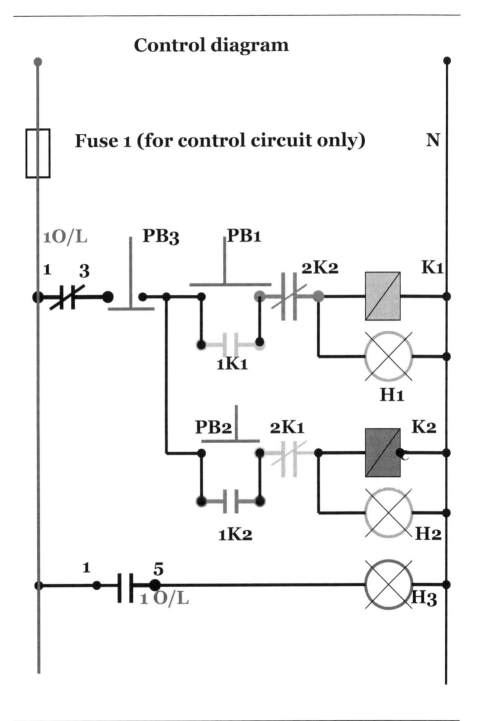

Fuse 1 (for control circuit only)

N

1O/L

PB3   PB1

1   3   2K2   K1

1K1

H1

PB2   2K1   K2

1K2   H2

1   5
1 O/L   H3

The significations of the symbols will be:

H1   Contactor K1 is ON

H2   Contactor K2 is ON

H3   Overload protection –alarm

Note: the outside line of the lamp H1/2/3 indicates the colour displayed when the light is ON status.

PB1 Push button motor rotation left

PB2 Push button motor rotation right

PB3 Push button to STOP the motor

K1 Contactor supplying the motor in sequence L1/L2/L3

1K1 NO (normally open) control contact located on the K1 contactor's coil

2K1 is the NC (normally closed) control contact located on the K1 contactor

K2 Contactor supplying the motor in sequence L2/L1/L3

1K2 NO (normally open) control contact located on the K2 contactor's coil

2K2 is the NC (normally closed) control contact located on the K2 contactor

Control diagram

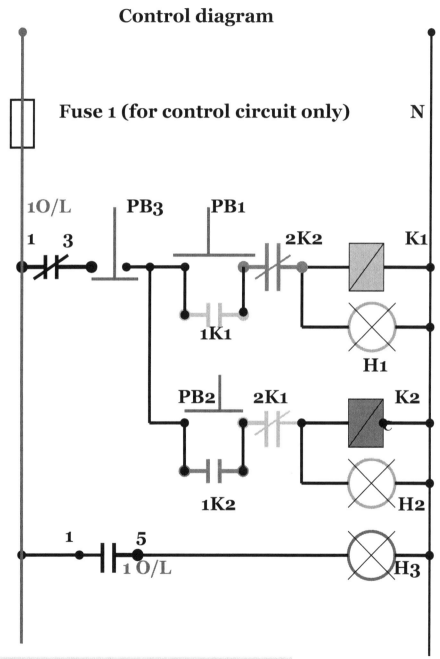

Fuse 1 (for control circuit only)   N

1O/L   PB3   PB1   2K2   K1

1   3   1K1   K2   H1

PB2   2K1   K2

1K2   H2

1   5
1 O/L   H3

There is no power to the motor.

Rotation to left push PB1 so K1 will connect and K2 will stay OFF

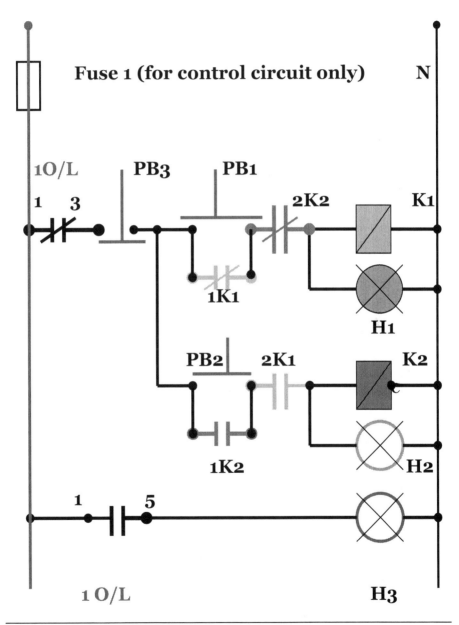

Fuse 1 (for control circuit only)   N

1O/L   PB3   PB1   2K2   K1

1   3   1K1

H1

PB2   2K1   K2

1K2   H2

1   5

1 O/L   H3

We push PB3 the diagram shown in page 117 will return to the status shown on page 116 so motor stops Let's now push PB2, causing the diagram to change into the status below; the motor will rotate to right

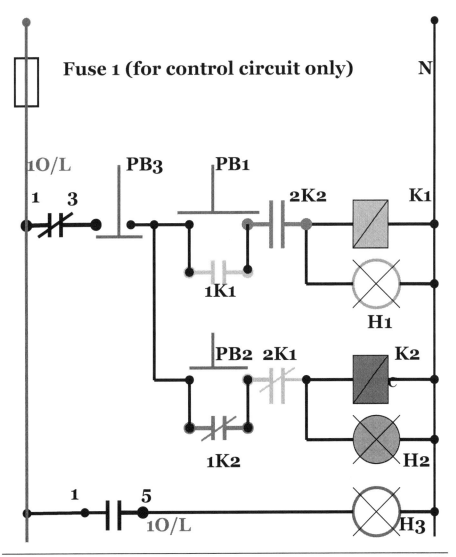

This representation will indicate the faulty status where there is an overload and the motor should be protected by the O/L relay (overload relay)

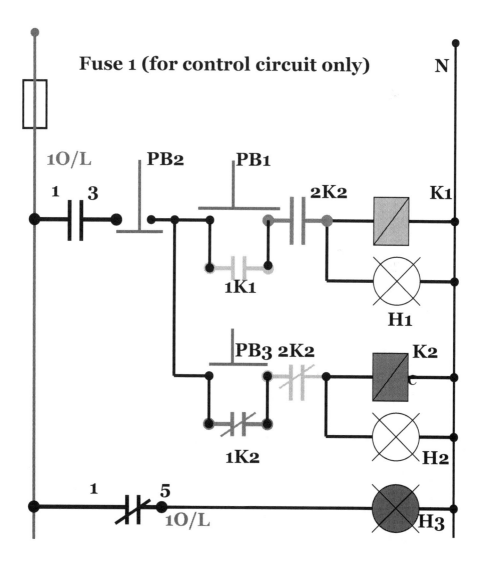

In case of an OVERLOAD, the relay O/L will work as described in page 111&112; the NO contact 1O/L will open and cut the power supply for one of the contactors (K1 or K2), simultaneously closing the NO contact 1O/L at 1-5 and supplying the H3 signal lamp with power. This will be illuminated as an alarm since an overload protection is being activated.

The status of the diagram will be:

H3 is ON; H1 and H2 is OFF

K1 coil de-energized

K2coil de-energized

H1 and H2 are OFF

As you can observe from the diagram, a fuse will be installed on the phase line only and neutral will go directly to the devices such as coils or indicating lamps. The hot wire (the phase) will go through the control's circuit. The neutral goes to the contactor's coils directly. *This sequence is very important and needs to be followed exactly as indicated.* Reversing the hot wire with neutral may create dangerous situations. Having the hot wire directly to the contactors for example will create a situation in this diagram where any accidental connections to grounding in places like these indicated in the diagram (X or Y)will determine the contactor to supply the motor with power without any command in this direction. The contactor K1 or K2 will be able to energize the motor because of the faulty situation. This is very dangerous situation. So there is very important how we connect the line and neutral to the control diagrams.

Looking to the diagram form the right it is easy to observe that any connection from point X or Y to Grounding(GND) will determine the contactor K1 or K2 to close his power contacts and this to supply the motor with power when in fact was an accidental situation. Here is the risky situation where devices will start because of failure of installations not because of the request of the process. This will make the diagram totally unsafe. So please pay attention to such details and most important follow the design requirements.

Wrong Control diagram

Neutral here (WRONG)

Line here (WRONG)

Fuse 1 (for control circuit only)

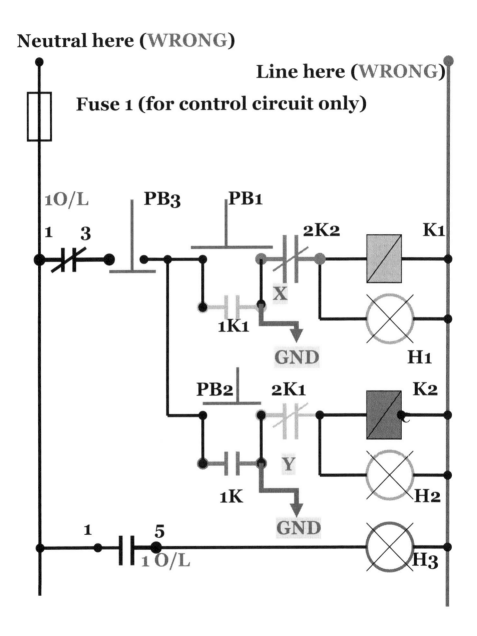

In our diagram we have three signal lamps. Most of the time these are located on the front of the panels/boxes in order to be visible. The lamps monitor the installation and provide important information regarding the status of the diagram and the status of the entire installation.

EXHAUST FAN POWER CIRCUIT 1Φ/3Φ & CONTROL DIAGRAM

Below is a diagram of a single-phase exhaust fan with power and control circuit:

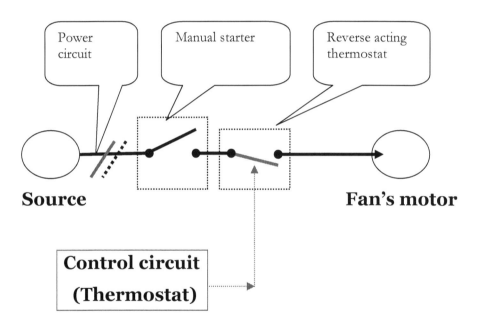

The control circuit includes a thermostat that will "read" the temperature in the room in which the fan is installed and it connects the power for the exhaust fan.

This next drawing illustrates the control diagram:

Electrical panel Manual starter

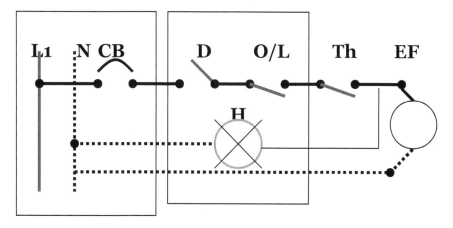

120v/60hz/1Φ

O/L: overload NC (normally closed) contact

Th: control/power contact NC (normally close)

D: power contact single-phase manual open in disconnect position. This is located in the manual starter.

H: Indication light/Electric Fan "ON"

CB: Circuit breaker located in the electrical panel. This circuit breaker is designed to protect the branch circuit. Both it and the manual starter should be ON in order to for the fan to be started. The "Th" represents the thermostat .This device is the control element of the diagram. Most of the time this device is installed by a mechanical tradesman (division 15).

As electricians, we work under division 16(26 form now) but sometimes our electrical drawings will request an installation of the EC (empty conduit) and electrical box for the control cable.

Please keep in mind that the thermostat should not be installed on the exterior walls of the building and that elevation is important as well. These two requirements are general requirements and must be applied unless otherwise specified on the drawings or specifications.

El. Panel   Manual Starter

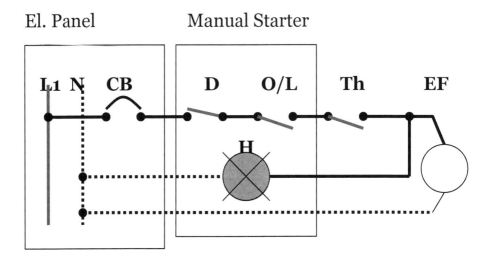

When the temperature reaches the prescribed values, the contact will open and stop the electrical fan EF.

If an overload occurs, the O/L contact will open and protect the motor.

Below is a manual starter diagram with the covers and overload section removed:

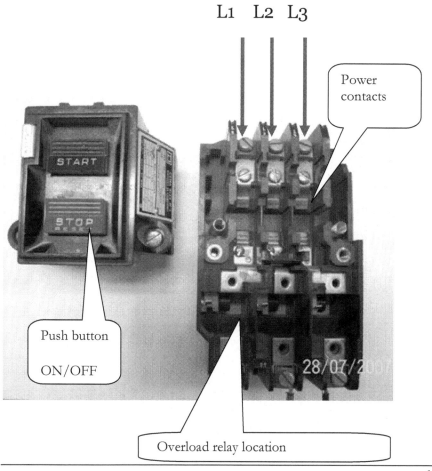

L1   L2   L3

Power contacts

Push button

ON/OFF

Overload relay location

"Overloads" installed (3 ea thermal units)

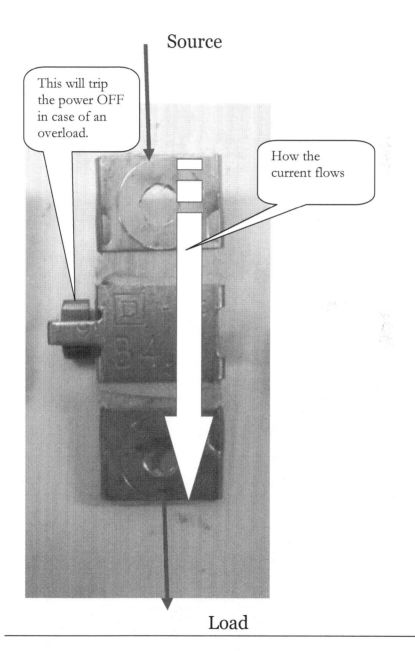

Thermal Unit (TU) detail

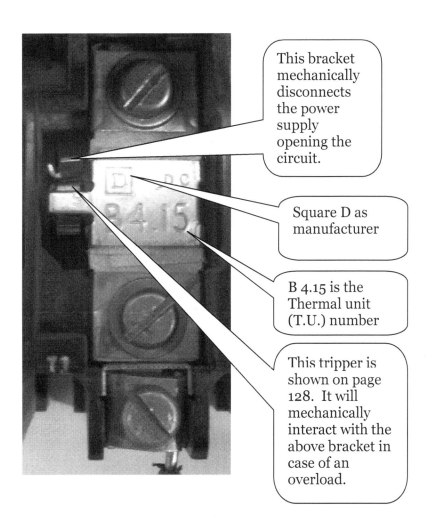

This bracket mechanically disconnects the power supply opening the circuit.

Square D as manufacturer

B 4.15 is the Thermal unit (T.U.) number

This tripper is shown on page 128. It will mechanically interact with the above bracket in case of an overload.

Each phase includes such devices. For clarity, only one phase is shown here.

This B 4.15 is the thermal unit number and will trip overloads in the range of: 2.52- 2.99 amperes, where 3T.U.(Thermal Units) are installed. The manufacturer will provide specifications for this.

THREE –PAHSE EXHAUST FAN: POWER AND CONTROL CIRCUIT

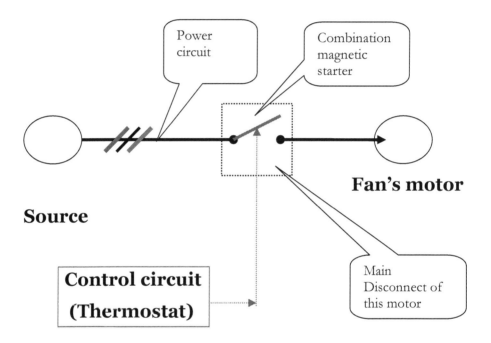

This next diagram will display the control circuit for an exhaust fan as a three-phase load.

COMBINATION MAGNETIC STARTER

208/3ph/60 hz

local disconnect

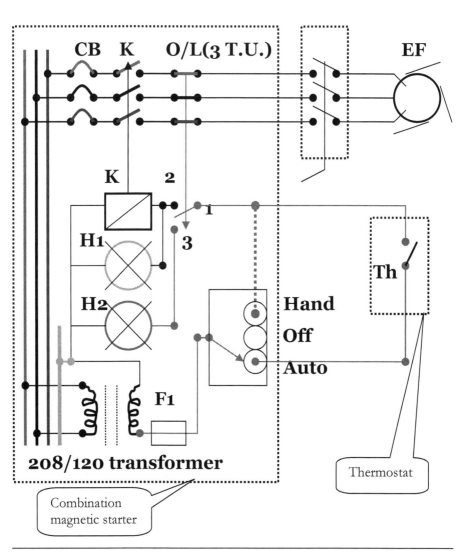

Combination magnetic starter

Thermostat

This diagram works as follows: there is a transformer installed in the magnetic starter for safety reason and for technical reasons as well.

The transformer provides 120 volt for the control circuit.

The main device triggering the control diagram is the thermostat.

This device will control the electrical fan (EF) in three stages: manual/off/auto

The electrical fan EF will run since the selector is on "AUTO "or "HAND" position.

On OFF position means no action will be taken since the electrical path to the coil is interrupted.

1. When the selector is on "AUTO" position, the fan will run since the thermostat contact is closed due to the high temperature. During this operation the fan supplies fresh air to the area so that the temperature will drop. When the

temperature drops, the thermostat's contact re-opens and the fan stops.

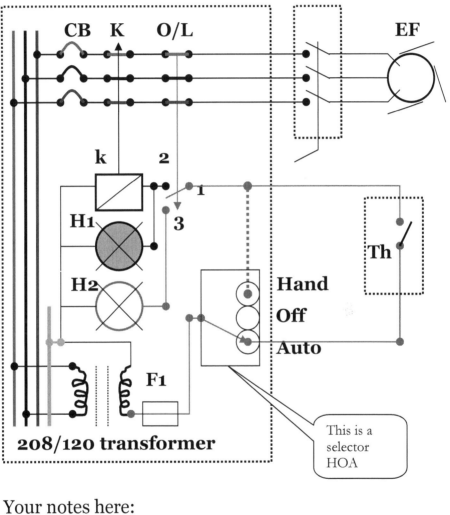

Your notes here:

----

2. When the selector is on "HAND" position, the fan will run whether the thermostat is requiring it to or not. As you can see in the diagram, in such a case, the thermostat has no effect on the diagram since the contactor's coil is energized through the "HAND" contact and the normally closed overload contact.

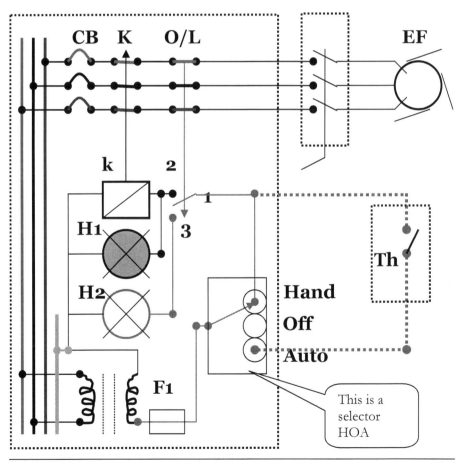

3. The "OFF "position of the selector will prepare the diagram for maintenance or intervention. The fan will not start in this stage

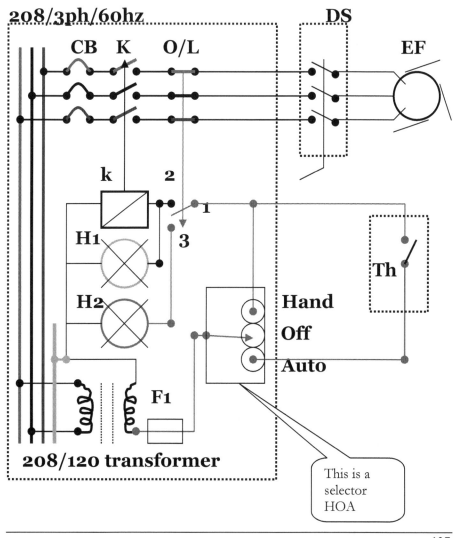

4. H1 will indicate the electrical fan's running status.

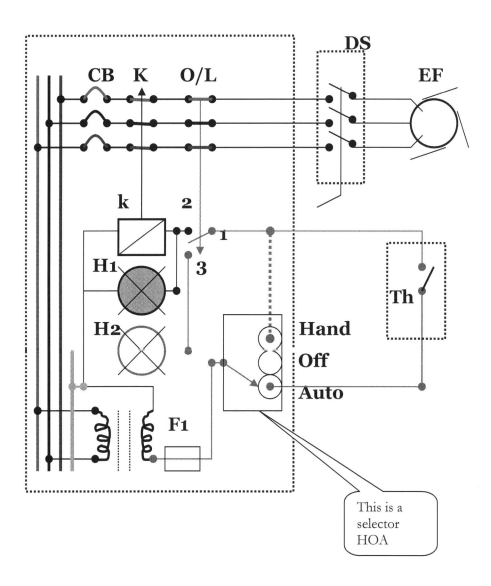

This is a selector HOA

5. H2 will indicate the overload status when the overload protection relay works and will open the normally closed contact 1-3 1O/L. At the same time, it will close contact 1O/L 1-2 energizing H2 to indicate the defect status.*

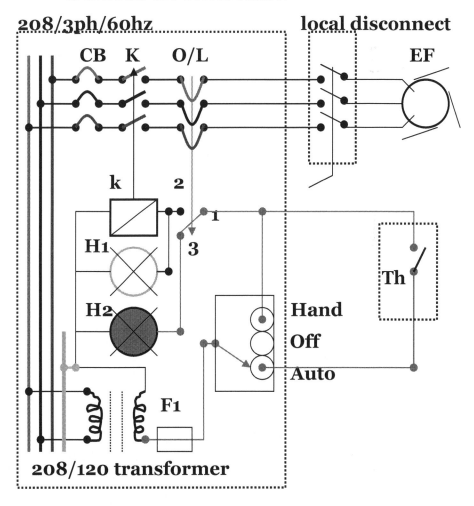

Also the contactor's coil will be losing its power supply so the fan will stop and the front panel will indicate that there is an "OVERLOAD" by way of a H2 light going "ON". The CB (circuit breaker) is 3-pole circuit breaker and rated as per the fan's current and voltage. Observe the F1 (fuse) for the control circuit.

Keep in mind the contactor and its coil are the two main parts of the starter.

     The contactor is provided with the power contacts in order to supply the load.

     The contactor's coil is the main element in the control diagram.

Attached to the contactor we may have auxiliary contacts (NO and NC) which are required for indications or different tasks. "H1" or the indication light will provide good information regarding the status of the equipment during testing or maintenance. The local disconnect is a safety requirement and inspectors will ask for this device. It does not have to be fusible but just a means of disconnecting means and needs to be installed somewhere visible to protect any maintenance personnel that might work on the fan for any reason. This disconnect provides the minimum safety requirements required since maintenance is needed. Here is an illustration of what the combination magnetic starter looks like:

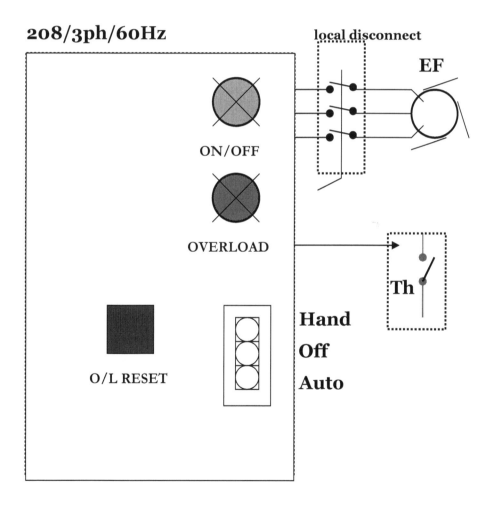

Most of the time the running status is indicated with the green light (H1) and the defect status is indicated by the H2 light. In some situations, the overload may just be a temporary event. In such a case, the overload will be indicated on the display. At that point, we need to verify which part of the installation is affected.

If no major defect is found, we can reset the overload contact by pressing the blue button. That will basically bring the contact 1-3 of 1O/L back into its normally close (NC) position. It will also bring the control circuit back to its initial status (prior to the indication of an overload [H2light]).

Please note: You might find that the colour codes vary in different installations.

Indication light "H1" and "SELECTOR" are located on the starter's cover.

In the following photo the Hand-Off-Auto is missing.

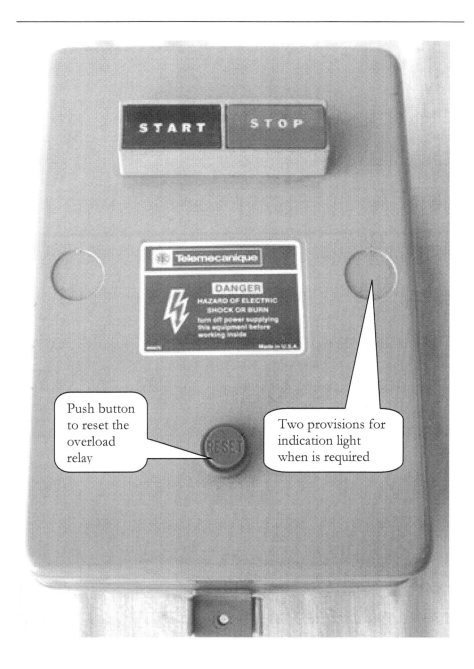

3-WAY CONTROL SWITCH FOR LIGHTING SYSTEM

This is designed to make lighting control and switch operations possible from two different locations.
This kind of switch is installed when there is a room with two entrances (doors) and there is a need to control the lighting in the room from switches located at the doors. The following diagram illustrates this situation:

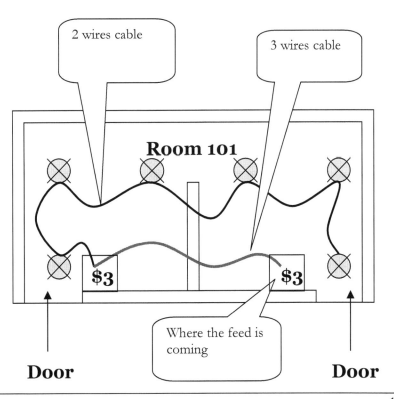

Lighting fixtures "ON"

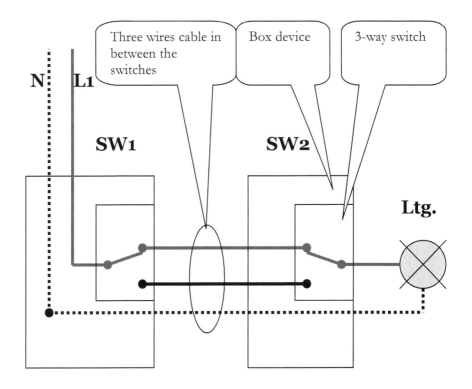

The 3-way switch has three poles and two positions.

It may have one terminal for grounding as well.

Between the switches we need to **run three wires cable** as is indicated in the diagram.

This kind of lighting control is designed for high frequency use at such projects as retirement homes.

The use of this control makes day-to-day activities more comfortable and easier.

Any of the switches will be able to turn OFF the lighting:

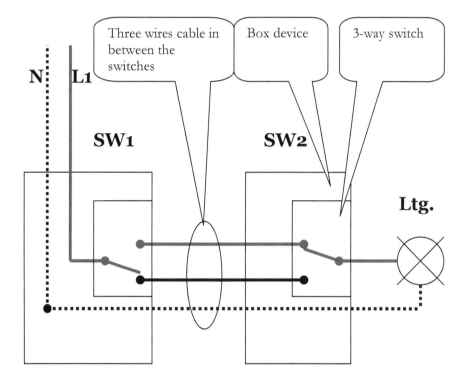

*SW 1 is "OFF ", SW2 remains on in the same initial position and the light goes "OFF." *

When SW 1 is "OFF", SW2 will change position and the light will go "ON."

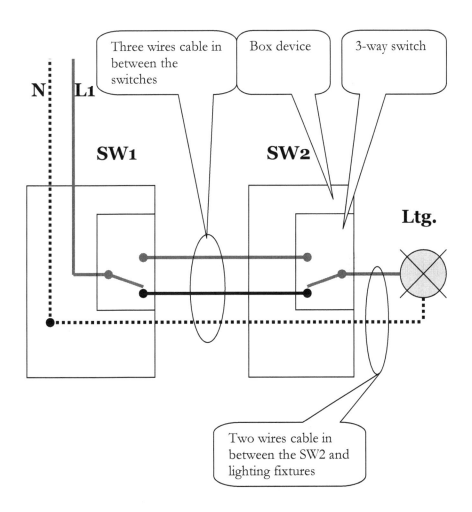

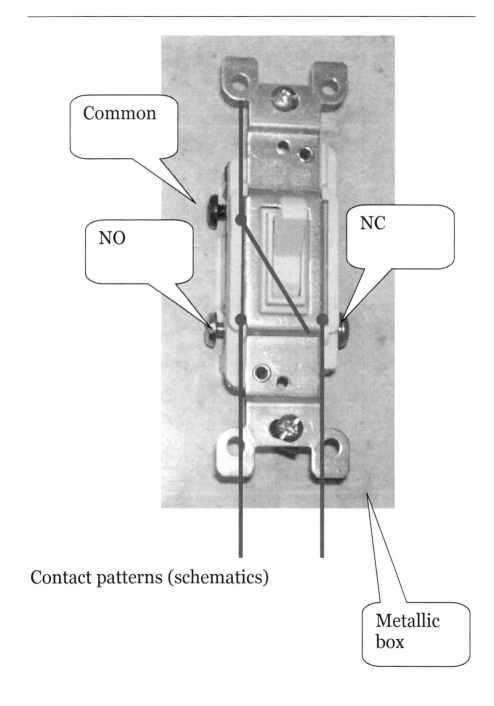

Common

NO

NC

Contact patterns (schematics)

Metallic box

2 conductor cable

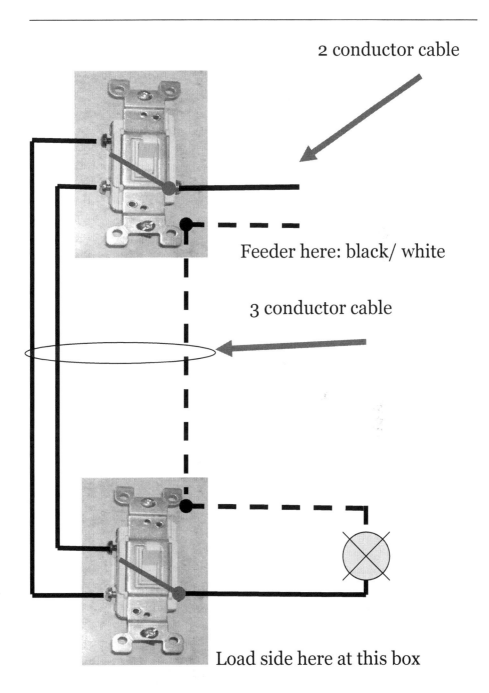

Feeder here: black/ white

3 conductor cable

Load side here at this box

4-WAY CONTROL SWITCH FOR LIGHTING SYSTEM

This is designed to make it possible for lighting switches to operate from three or more different locations.

This kind of switch is installed when there is a room with more than two entrances (doors) and there is a need to control the lighting in the room from switches located at all the doors/entrances.

The diagram below illustrates this situation:

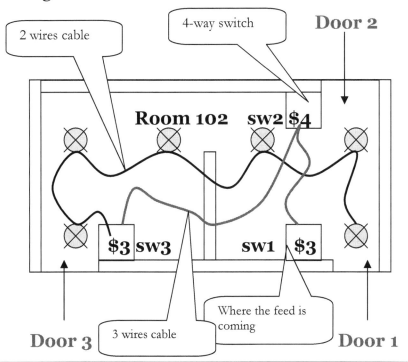

The next diagram shows a 4-way switch connection:

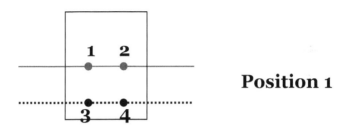

Position 1

Position 1 will make these contacts: 1&2 and 3&4

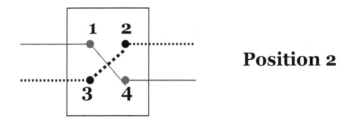

Position 2

In position 2 will make these contacts:

1&4 and 3&2

As you can see there are 3 way switches at the extremities and 4 way switches in between. This is for 3 locations control. You need 5 points of control installed, including 2 more at the 4-way switch between the 3 way switches.

So the complete drawing for lighting control requiring at least three point of control looks like this:

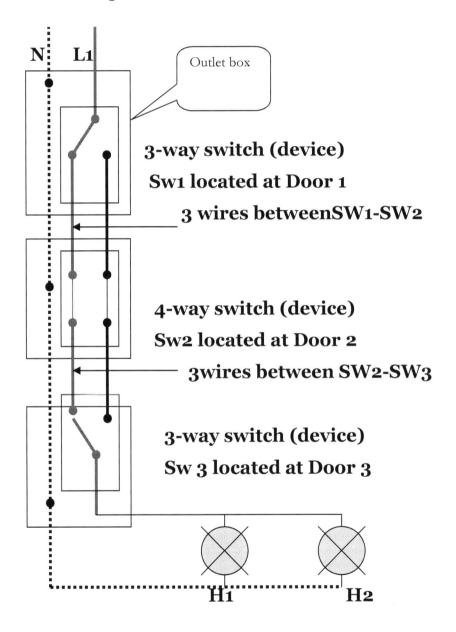

Outlet box

3-way switch (device)

Sw1 located at Door 1

3 wires betweenSW1-SW2

4-way switch (device)

Sw2 located at Door 2

3wires between SW2-SW3

3-way switch (device)

Sw 3 located at Door 3

ELECTRICAL DEVICES INTERCONNECTED WITH FIRE ALARM SYSTEM

Technological advances have been made in all kinds of equipment, including fire alarm systems which have been optimized and upgraded.

This means it is possible to create large applications with accurate indications and alarm signals.

Following a fire, there will be an alarm and signal to the main fire alarm control panel (FACP).

There are devices that are interlocked with the fire alarm system in order to avoid dangerous situations during or after a fire.

For example, the supply fans that bring fresh air into the area must be automatically disconnected in order to prevent bringing fresh air into an area where a fire has occurred.

As we all know, fire consumes air (oxygen) and, therefore, decreasing the air in an area affected by fire will help extinguish it.

However, adding more air and oxygen will create a very dangerous situation in which a fire may increase in intensity and become very hard to control.

In order to recognize such situations, we need to understand how the alarm or signal contact from a fire alarm system is integrated in the control diagram for such equipment.

As we learn about this, we will integrate the discussions and diagrams from previous chapters that referred to the power and control circuit for supply air fans.

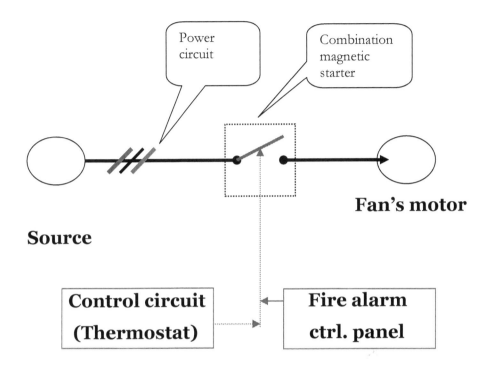

The thermostat and fire alarm devices control the starter of the electrical fan (EF).

So here we have not the control circuit only but the interconnection between the power circuits –control circuit- fire alarm system. Such interconnection is imposed by the request to stop electrical fan supplying fresh air in event of a fire.

The next diagram displays the control circuit for this fan in relation to the fire alarm system diagram.

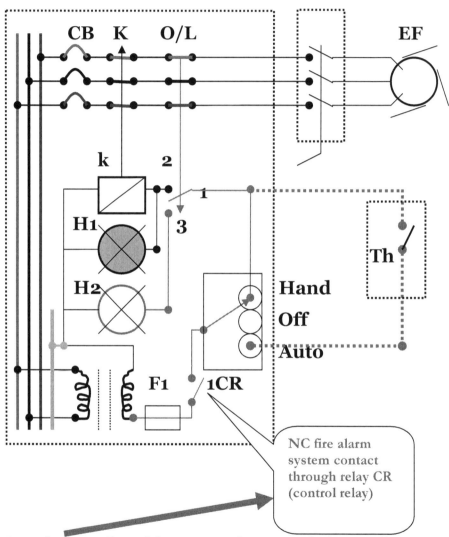

NC fire alarm system contact through relay CR (control relay)

We observe that this contact is NC and since a fire event that occurs will trigger the control relay (CR) to change its status from NC to NO this will be the equivalent of a circuit opening so that the contactor for power "K" will never connect the power to the load. Obviously if this motor is running and a fire occurs then contactor K will lose power on his coil and will disconnect the power from the load. As a result no fresh air will enter the area affected by the blaze.

Not only the fans are interlocked with fire alarm system but also devices that are part of the system that will provide the access (as door openers) or elevators.

Having a fire event the doors locked with magnetic locks should be released so there is a free way for people to evacuate such areas and the elevators are going automatically to designated locations so the humans will not be locked in elevators and they will be directed at specific locations (This is "homing "or elevator call function).

These are aspects that you need to know and be aware of them as being part of our work as electricians.

Fire alarm systems are designed to have two kinds of circuits:

1. Initiation circuit (that will trigger the alarm)

2. Notification circuit (that will "notify" with strobes/horns when a fire event is in progress)

The ***Initiation circuit*** is composed of two wires class "A" in loop connecting next main devices: smoke detectors, heat detectors, duct detectors and pull stations. Also this circuit includes devices such as:

- Control relays (CR),
- Line Isolator modules (IM) and
- Input/output (CT1 or CT2) modules

These devices are part of the loop and will change their status based on the current on the loop.

For example, when one of the detectors or pull stations initiates the alarm, the CR will change the status of his contacts. The CR's contacts are show below.

Schematic of the contacts will be:

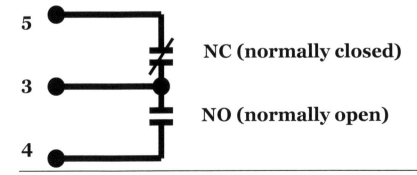

NC will open and NO will close as a result of a fire event.

They are integrated (see 1CR/NC contact in control diagram page 156) in the devices' control diagram and will disconnect the power supply if alarm is initated.

This is the final result and will achieve the main goal which is to make installation dependable for the fire alarm system and SAFE.

Devices as control relays (CR) or modules (CT1 or CT2) are most of the time addressable and they are part of the initiation loop being able to "observe" any event in the loop (activation of a smoke detector or pull station).

The control relay for example is most of the time designed to be interconnected with the make-up units; air-conditioning units; electrical fans and others mechanical devices. They have contacts integrated with the control diagram of the equipment they control.

The relay is connected to the loop with two wire system and having an address will be programmed to change the status of his contacts in that way that stops the equipment he is controlling through his contact when a fire alarm will occur.

CT1 or CT 2 is a device that is part of the initiation loop as well. They have been designed to integrate in the loop devices as:

- pressure switches(PS)
- Flow Switches(FS)
- Micro switches indicating the open/close status of the mechanical valves (V).

These devices are the control devices located in any sprinkler system.

Sprinkler system is the one will work during the fire event and is been designed to extinguish the fire in the area where is installed.This system needs to be ready to work in any moment. For this reason devices like valves, pressure switches or flow switches part of this system needs to be monitored or /and supervised by the fire alarm system.

CT's will supervise these devices and will notify when the pressure in the sprinkles system is low or a valve that normally should be closed are open.

There is a variety of devices on the market and the way they are working seems to the almost the same. In fact it is.

Most of the system I did install was Edwards and Mircom and I been impressed by these devices and about the service we did receive from Edwards and Mircom through their managers and technicians.

Valve supervision by fire alarm system using CT1 device

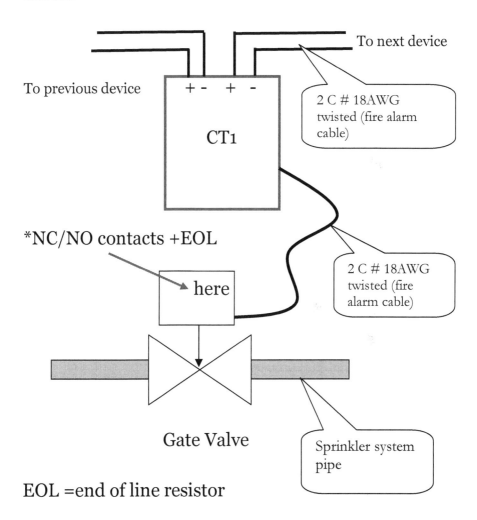

To next device

To previous device

+ - + -

CT1

2 C # 18AWG twisted (fire alarm cable)

*NC/NO contacts +EOL

here

2 C # 18AWG twisted (fire alarm cable)

Gate Valve

Sprinkler system pipe

EOL =end of line resistor

Valve and system pressure supervisory by fire alarm system using CT 2 device:

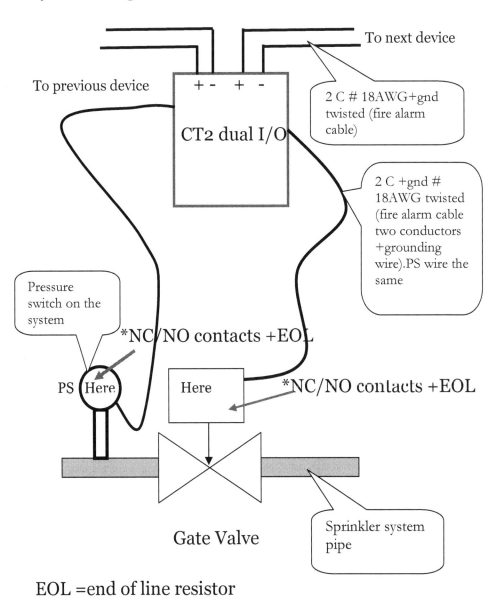

EOL =end of line resistor

The next block diagram will show how the system is working as **Initiation circuit**.

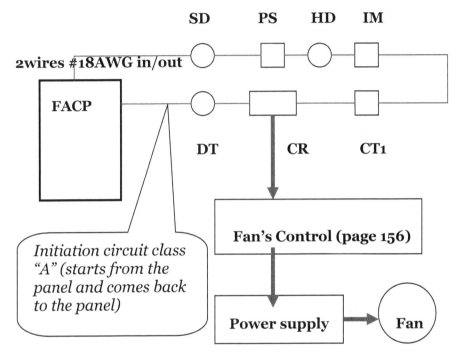

This system is using # 18 AWG wiring because of the specification of the loads (electronic devices-low current for operation) .On these circuits currents will flow at less than 1 ampere (mA) since the devices are mostly electronic. This results in a limitation on numbers of devices on the loop. Using Edwards's parts, you can have about 250 devices in total on the loop.

This is a large number of devices. The devices on the loop are addressable and able to work on 2 wires only.

For further clarification, the interpretation of the symbols used can be found below:

FACP= fire alarm control panel

SD=smoke detector

HD=heat detector

DD=smoke duct detector

PS=pull station

IM= Line isolation module

CR=control relay

CT=current input/output module (single or dual)

Notification Circuit This circuit includes devices that are able to give notification in case of a fire. This alerts all humans located in the area where an emergency is happening. As a result people will leave the building or location. The manner of evacuation is based on the procedure established by the management of the building. The specified devices attached to the notification circuit are:

- horns,
- strobes,
- combination horns/strobes,
- Bells and speakers.

Devices such as these consume more energy than devices located on an initiation circuit so the power supply for them should be designed accordingly.

They are working in DC power so the system will be provided with power supply sources for direct current (DC) in fact batteries to carry the loads for certain amount of time since during the fire the power supply is at high risk.

Batteries are available and chargers able to make the system viable during a normal power system failure. In most cases this type of circuit will be installed as class

"B" having an end of line resistor (EOL) as the last device.

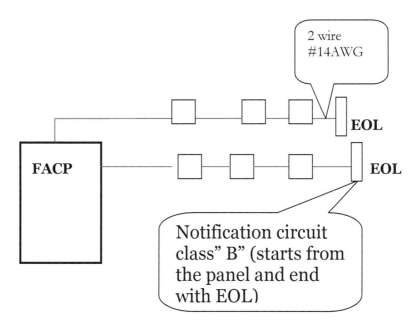

This system is using # 14 AWG wiring because of the specification of the loads. No more than 8 devices for the Edwards System.

You may find different but is better to verify the manufacturer specifications

Mircom System will accept more devices depending on the DC power source. They require no more than 1.2 amperes for the notification circuit.

Most of the devices will require about 0.035 amperes. (This is for the mini-horns)

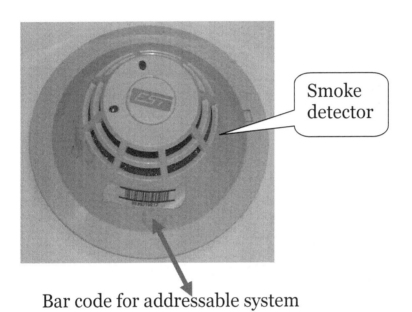

Smoke detector

Bar code for addressable system

Bar code for addressable system (keep this for programming)

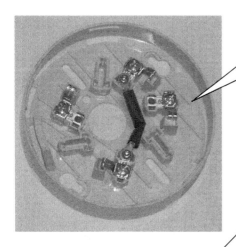

Detector Base

Detector's terminals to fit the detector's base.

Heat Detector

Your notes here:

....
....

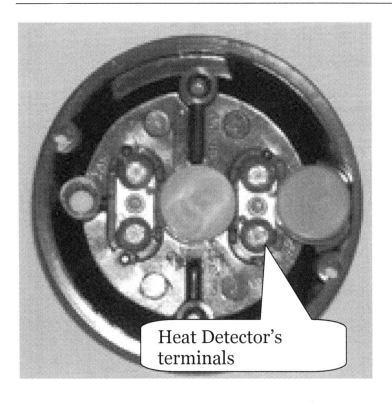

Heat Detector's terminals

Your notes here:

....
....
....
....
....
....
....
....
....
....
....
....

Glass rod (will break when the pull station is activated)

Pull Station

Install the resistor here

EOL Plate /back view

Mini-horn

Mini-Horn /Strobe

For more details and information regarding fire alarm systems, please read my book: Electrician's Book: Fire Alarm System Installation.

FINAL WORDS

We now have something in common having journeyed together through the one hundred and seventy + pages of this book. We share a passion and fascination for this field and career. I am happy to have been able to pass on my experiences and knowledge to you so that you could learn! Congratulations for reading and learning from this book. Please don't forget to regularly review the concepts you've learned about in these pages.

Whether you are an apprentice or electrician, I wish you well in reaching your goal of working successfully and securing your future!

Cornel Barbu

Toronto, July 14, 2007

Made in the USA
Lexington, KY
29 May 2016